FOOD SCIENCE AND TECHNOLOGY



# FOOD LABELING

## FDA PROTECTIONS AND COUNTRY-OF-ORIGIN LABELS

# FOOD SCIENCE AND TECHNOLOGY

Additional books in this series can be found on Nova's website under the Series tab.

Additional E-books in this series can be found on Nova's website under the E-book tab.

# GOVERNMENT PROCEDURES AND OPERATIONS

Additional books in this series can be found on Nova's website under the Series tab.

Additional E-books in this series can be found on Nova's website under the E-book tab.

FOOD SCIENCE AND TECHNOLOGY

# FOOD LABELING

## FDA PROTECTIONS AND COUNTRY-OF-ORIGIN LABELS

JOSEPH K. SCOTT
AND
MARY I. HAYES
EDITORS

Nova Science Publishers, Inc.
*New York*

For permission to use material from this book please contact us:
Telephone 631-231-7269; Fax 631-231-8175
Web Site: http://www.novapublishers.com

## NOTICE TO THE READER



**LIBRARY OF CONGRESS CATALOGING-IN-PUBLICATION DATA**

ISBN: 978-1-61942-759-4

*Published by Nova Science Publishers, Inc. † New York*

# CONTENTS

# PREFACE

Consumers increasingly seek information on food labels to help them make healthy food choices, and they rely on federal agencies to ensure that such information is truthful. Food companies may use health- and/or nutrient-related claims on food labels to appeal to consumers, distinguish their products from their competitors', and otherwise increase their sales. However, when those claims are false or misleading, consumers' efforts to select healthy food may be undermined. This book provides an overview of FDA protocols to reassess its approach in detecting and protecting consumers from false or misleading claims on food labeling.

Chapter 1 - The Food and Drug Administration (FDA) oversees federal requirements to prohibit false or misleading food labels; the Federal Trade Commission enforces the prohibition against false or misleading advertising. By statute, health claims on food labels must have significant scientific agreement, but in 2002, in response to a court decision, FDA decided to allow qualified health claims with less scientific support. Structure/function claims refer to a food's effect on body structure or function and are also used on food. Congress directed GAO to study FDA's implementation of qualified health claims for food. GAO examined (1) the results of FDA's efforts to allow the use of qualified health claims and oversight of these claims and (2) consumers' understanding of the claims. GAO also examined FDA's oversight of structure/function claims. GAO reviewed FDA documents and consumer studies and interviewed stakeholders from health, medical, industry, and consumer groups.

Chapter 2 - Many retail food stores are now required to inform consumers about the country of origin of fresh fruits and vegetables, seafood, peanuts, pecans, macadamia nuts, ginseng, and ground and muscle cuts of beef, pork, lamb, chicken, and goat. The rules are required by the 2002 farm bill (P.L. 107- 171) as amended by the 2008 farm bill (P.L. 110-246). Other U.S. laws have required such labeling, but only for imported food products already pre-packaged for consumers.

In: Food Labeling: FDA Protections...
Editors: J. K. Scott and M. I. Hayes
ISBN: 978-1-61942-759-4


*Chapter 1*

# Food Labeling: FDA Needs to Reassess Its Approach to Protecting Consumers from False or Misleading Claims

*United States Government Accountability Office*

## Why GAO Did This Study

The Food and Drug Administration (FDA) oversees federal requirements to prohibit false or misleading food labels; the Federal Trade Commission enforces the prohibition against false or misleading advertising. By statute, health claims on food labels must have significant scientific agreement, but in 2002, in response to a court decision, FDA decided to allow qualified health claims with less scientific support. Structure/function claims refer to a food's effect on body structure or function and are also used on food. Congress directed GAO to study FDA's implementation of qualified health claims for food. GAO examined (1) the results of FDA's efforts to allow the use of qualified health claims and oversight of these claims and (2) consumers' understanding of the claims. GAO also examined FDA's oversight of structure/function claims. GAO reviewed FDA documents and consumer studies and interviewed stakeholders from health, medical, industry, and consumer groups.

## WHAT GAO RECOMMENDS

GAO recommends FDA identify and request from Congress authorities to access companies' evidence for potentially false or misleading structure/function claims on food to establish scientific support, provide guidance to industry on the evidence it needs to support such claims, and provide direction to FDA inspectors to help identify claims for further review. FDA generally agreed with the first two recommendations but found the third to be impractical; GAO clarified that recommendation.

## WHAT GAO FOUND

FDA's efforts to protect consumers from false or misleading claims are conducted in a complex and challenging legal and regulatory environment. From 2002, when FDA announced its decision to allow qualified health claims on food labels—following a court decision involving its authority to regulate dietary supplements—through September 2010, FDA received 16 petitions from companies proposing 60 claims on food labels. After reviewing the scientific evidence presented in the petitions, FDA determined that there was enough credible scientific evidence for the agency to allow the use of 12 qualified health claims, by modifying language to qualify the claims and characterize in detail the strengths and limitations of the scientific support for those claims. In overseeing qualified health claims for food labels, FDA has issued two warning letters to food companies—both in 2010—for citing health benefits that were not in the allowed qualified health claims or supported by scientific evidence.

Research showed, and stakeholders indicated, that consumers find it difficult to understand the differences between health claims with significant scientific agreement and the lower level of scientific support for qualified health claims. Research also showed that consumers find it difficult to distinguish among the many different types of claims on food labels, including health claims, qualified health claims, and structure/function claims.

FDA data indicate that companies now minimally use qualified health claims on foods but more widely use structure/function claims to convey their foods' health benefits. Companies' use of structure/function claims is subject to the general statutory requirement that labeling not be false or misleading. However, FDA has not given companies guidance on the scientific support needed to prevent false or misleading information for a structure/function claim for food or given its inspectors instructions for identifying potentially false or misleading information in such claims when examining food labels as part of food facility compliance inspections. Even if FDA were to provide such guidance, structure/function claims pose a serious oversight dilemma for the agency. That is because FDA—unlike the Federal Trade Commission (FTC), which can require companies to submit any relevant evidence as part of an investigation of whether claims are substantiated—does not have the ability to compel companies to turn over their substantiation documents. GAO's work indicates that FDA's efforts to meet that burden are hampered by the lack of access to the evidence that a company relies on to make such a claim. In particular, while FDA may ask a company to provide its scientific support for a claim, FDA does not have express legal authority to compel the company to provide such information. FTC, on the other hand, which is responsible for protecting consumers from false advertising generally, has the authority to compel companies to provide the support. FTC officials said that the Commission would have difficulty taking enforcement actions against companies for alleged false structure/function claims on food labels and in advertisements without access to companies' proprietary market and scientific research.

## ABBREVIATIONS

| | |
|---|---|
| CFSAN | Center for Food Safety and Applied Nutrition |
| EU | European Union |
| FDA | Food and Drug Administration |
| FLAPS | Food Label and Package Survey |
| FTC | Federal Trade Commission |
| FTE | full-time equivalent |
| G | gram |
| HHS | Department of Health and Human Services |
| IOM | Institute of Medicine |
| LDL | low-density lipoprotein |
| mg | milligram |
| NLEA | Nutrition Labeling and Education Act |
| RACC | referenced amount customarily consumed |

January 14, 2011

The Honorable Daniel K. Inouye
The Honorable Thad Cochran
United States Senate

The Honorable Harold Rogers
Chairman
The Honorable Norman D. Dicks
Ranking Member
Committee on Appropriations
House of Representatives

Consumers increasingly seek information on food labels to help them make healthy food choices, and they rely on federal agencies to ensure that such information is truthful. Food companies may use health- and/or nutrient-related claims on food labels to appeal to consumers, distinguish their products from their competitors', and otherwise increase their sales. However, when those claims are false or misleading, consumers' efforts to select healthy food may be undermined. Under federal statutes and regulations, health claims on food must either (1) be authorized by FDA upon a determination that "significant scientific agreement" exists among qualified experts that the totality of publicly available scientific evidence supports the claims or (2) be based on an authoritative statement of a scientific body of the federal government or the National Academy of Sciences.

Through separate statutes, the Food and Drug Administration (FDA) and the Federal Trade Commission (FTC) share jurisdiction over health- and nutrient-related claims made by food manufacturers. FDA oversees federal food labeling requirements that prohibit, among other things, food labeling that is false or misleading; FTC oversees federal consumer protection requirements that prohibit, among other things, deceptive acts or practices in advertising, including food advertising. Under a memorandum of understanding, the two agencies agreed that FDA has primary responsibility for food labeling and FTC has primary responsibility for food advertising. FTC has recently emphasized that health claims on food must be adequately substantiated and presented in a manner that is truthful and not deceptive. FDA regulations allow companies to use an authorized health claim on a label or petition FDA for authorization of a new health claim. To ensure health claims have the proper support, FDA analyzes and reviews research on nutrition and the role of food in maintaining health.

FDA categorizes health- and nutrient-related claims on food labels as follows:

- *Health claims* characterize the relationship of any substance to a disease or health-related condition (e.g., diets low in sodium may reduce the risk of high blood pressure, a disease associated with many factors).
- *Structure/function claims* describe the role of, or characterize the mechanism by which, a nutrient affects a body structure or function (e.g., calcium helps build strong bones).
- *Nutrient content claims* characterize the level of a nutrient in a food (e.g., good source of vitamin C).

Qualified health claims are health claims that characterize the relationship of a food component to a disease or health-related condition, as modified with a disclaimer or otherwise qualified by FDA to characterize the strengths and limitations of the scientific support to address the claim's potentially misleading nature. For example, the following is a qualified health claim allowed by FDA characterizing the relationship between the consumption of tomatoes and the risk of gastric cancer:[1] "Four studies did not show that tomato intake reduces the risk of gastric cancer, but three studies suggest that tomato intake may reduce this risk. Based on these studies, FDA concludes that it is unlikely that tomatoes reduce the risk of gastric cancer." Qualified health claims for food are not provided for in statute. Instead, they came about as a consequence of FDA's decisions following a 1999 federal appeals court ruling (*Pearson v. Shalala*). The court held that the First Amendment does not permit FDA to prohibit a potentially misleading health claim on a dietary supplement label unless FDA considers whether qualifying language on the label could negate the potentially misleading nature of the claim. (Dietary supplements are products that contain certain dietary ingredients such as vitamins, minerals, or herbs, intended to supplement the diet, and are labeled as dietary supplements.)

After the court ruling, in 2000, FDA announced an interim enforcement strategy to allow companies to petition the agency for qualified health claims on dietary supplements and established criteria for the agency to consider in exercising its enforcement discretion to *not* take actions. In 2002, FDA announced its decision to expand this approach to include qualified health claims on conventional food as well. The next year, FDA issued guidance to industry on procedures for petitioning the agency for qualified health claims for food. Once FDA announces, on the basis of its review of scientific evidence, that it intends to consider exercising its enforcement discretion for the use of a qualified health claim, any food company that meets the same criteria with the same food may use the claim.

In 2008, we reported that stakeholders from health, consumer, and medical groups advocated eliminating qualified health claims on food labels because they confused and misled consumers and might encourage consumption of foods with few or no health benefits.[2]

The Subcommittees on Agriculture, Rural Development, Food and Drug Administration, and Related Agencies, Senate and House Committees on Appropriations, directed that we study issues regarding FDA's implementation of qualified health claims on food.[3] Specifically, this report examines (1) the results of FDA's efforts to allow the use of qualified health claims on food, and its oversight of these claims and (2) what is known about consumers' understanding of qualified health claims on food. In addition, the report examines industry's use of structure/function claims on food and FDA's oversight of these claims. Appendix I provides updates to information we reported in 2008 on FDA's implementation and administration of health claims, including qualified health claims, in response to *Pearson v. Shalala*.

The term "food" is used throughout this report to mean "conventional food," not to include dietary supplements or animal feed. To determine the results of FDA's efforts to allow the use of qualified health claims on food, we assessed FDA data and documentation, including industry petitions and FDA guidance; reviewed research on companies' use of health- and nutrient-related claims on foods; and obtained the views of industry, FDA, and others on the advantages and disadvantages of using qualified health claims. To determine FDA's oversight of qualified health claims, we analyzed FDA guidance and related documents on the enforcement of food labeling claims, analyzed warning letters and other FDA actions, and compared and contrasted FDA's and FTC's responsibilities and authorities with respect to food labeling and advertising claims. To determine what is known about consumers' understanding of qualified health claims on conventional food labels, we reviewed research by FDA and others on consumers' understanding of such claims and interviewed stakeholders— researchers and industry and consumer, medical, and health groups we identified with input from the National Academies' Institute of Medicine, and FDA officials—for their views on what the research demonstrates about consumer perception. To examine industry's use of structure/function claims on food labels and FDA's oversight of these claims, we assessed FDA data, documentation, and studies on industry's use of these claims, and efforts to oversee their use. We also discussed the use of structure/function claims with Canadian and European Union officials. We also examined FTC's actions to correct structure/function claims it found to be deceptive in food advertising. We assessed the quality of the studies, including how the data they contained were initially developed and analyzed, and found those data to be of sufficient quality and reliability for the purposes used.

We conducted this performance audit from January 2010 through December 2010 in accordance with generally accepted government auditing standards. Those standards require that we plan and perform the audit to obtain sufficient, appropriate evidence to provide a reasonable basis for our findings and conclusions based on our audit objectives. We believe that the evidence obtained provides a reasonable basis for our findings and conclusions based on our audit objectives. Appendix II contains more detailed information on our objectives, scope, and methodology.

# BACKGROUND

The following sections discuss (1) the statutes, regulations, and guidance that underpin FDA's framework for overseeing claims on food labels and (2) FDA's organization for overseeing food labels.

## FDA's Framework for Overseeing Claims on Food Labels

FDA's framework for overseeing food labeling is based on statutes and FDA regulations and guidance. Specifically, the Federal Food, Drug, and Cosmetic Act prohibits the misbranding of food, which includes, among other things, food labeling that is false or misleading. The Nutrition Labeling and Education Act of 1990 (NLEA), amended the Federal Food, Drug, and Cosmetic Act to require that health claims for food labels be authorized by FDA following a determination by FDA that "significant scientific agreement" exists among qualified experts that the totality of publicly available scientific evidence supports the claims. According to FDA documents, the primary goals of NLEA were to (1) make nutrition information available to assist consumers in selecting foods that could lead to healthier diets; (2) eliminate consumer confusion by establishing definitions for nutrient content claims that are consistent and that consumers could rely on; (3) help consumers maintain healthy dietary practices and protect them from unfounded health claims, so a health claim used on a product would be one that consumers could rely on to give them truthful and not misleading information; and (4) encourage product innovation by developing and marketing nutritionally improved food.

FDA took several actions in response to the *Pearson v. Shalala* ruling, such as issuing guidance on how companies should submit petitions to FDA and how FDA would review the scientific evidence and exercise its enforcement discretion. (See app. I for information on FDA's implementation and administration of health claims in response to *Pearson v. Shalala.*) In 2003, FDA issued interim guidance to industry for submitting petitions for qualified health claims on food labels and for reviewing the scientific evidence for those claims, and explained its intention to consider exercising its enforcement discretion to *not* take enforcement actions against companies that use qualified health claims as FDA specified. In 2006 FDA issued additional guidance to industry that described how the agency would review the scientific evidence industry submitted in support of the proposed qualified health claims. FDA specified that a petition for qualified health claims should, among other things, (1) identify the relationship between the substance and the disease in the United States and demonstrate that the substance is a food, a food ingredient, or food component that has been shown to be safe and lawful at levels necessary to justify a claim; (2) propose a claim—a statement that describes the benefit of the food substance—and (3) present scientific evidence that supports the claim, including copies of computer literature searches on studies and all research articles supporting the claim. Petitions may be submitted by individuals, food manufacturing companies, food distributors, trade or professional associations, or a combination of these. For this report, we generally refer to petitioners as companies.

As specified in its 2003 and 2006 guidance, FDA conveys its determinations to allow qualified health claims and its intention to exercise enforcement discretion with respect to certain requirements of the Federal Food, Drug, and Cosmetic Act to petitioners. In letters to the companies announcing its intention to exercise enforcement discretion, FDA stipulates the wording of the qualified health claims and any conditions on their use. FDA also alerts petitioners that the claims must meet the general requirements for FDA's labeling regulations for health claims, *except* for the requirement that the evidence for the claim meet the significant scientific agreement standard and that the claim be made in accordance with an authorizing regulation. The qualified health claims and all conditions regarding their use are posted on

FDA's Web site to alert all food companies that the claims are available for use without further FDA review or approval. When FDA decides to deny a proposed claim, it describes its review of the scientific evidence and the rationale for its decision in correspondence to the petitioners and posts the letters conveying those denials on its Web site. In addition, in 2009 FDA issued final guidance for industry describing its process for the scientific review of health claims, including qualified health claims, in which it stated that it would reevaluate health claims and qualified health claims as new scientific evidence emerged.

As with health claims, companies also use structure/function claims to convey the health benefits of their foods or dietary supplements. The Dietary Supplement Health and Education Act of 1994 established special requirements for structure/function claims on dietary supplement labels. Specifically, under the act, if a dietary supplement label includes a structure/function claim, it must have a disclaimer stating, "This statement has not been evaluated by the Food and Drug Administration. This product is not intended to diagnose, treat, cure or prevent any disease." Products intended for use in the diagnosis, mitigation, treatment, cure, or prevention of a disease are considered drugs under the Federal Food, Drug, and Cosmetic Act. In addition, a structure/function claim on a dietary supplement label must have substantiation and either (1) claim a benefit related to a classical nutrient deficiency disease and disclose the prevalence of such disease in the United States, (2) describe the role of a nutrient or dietary ingredient intended to affect the structure or function in humans, (3) characterize the documented mechanism by which a nutrient or dietary ingredient acts to maintain such structure or function, or (4) describe general well-being from consumption of a nutrient or dietary ingredient.[4] These requirements do not apply to structure/function claims on food labels. However, the use of such claims for food is subject to the Federal Food, Drug, and Cosmetic Act's requirement that labeling *not* be false or misleading, and FDA has not issued guidance to companies regarding the proper use of structure/function claims on food labels.

Appendix III provides information on the different types of health- and nutrient-related claims that may be found on food labels. Appendix IV presents the 12 health claims for food that have significant scientific agreement and 5 health claims for food that are based on authoritative statements.

## FDA's Organization for Overseeing Food Labels

Within FDA, food labeling is under the jurisdiction of the Office of Foods' Center for Food Safety and Applied Nutrition (CFSAN). Within CFSAN, the Office of Nutrition, Labeling and Dietary Supplements is responsible for reviewing qualified health claim petitions, and the Office of Regulation, Policy, and Social Sciences conducts consumer research on health claims and qualified health claims. In addition, for more than 30 years, FDA has conducted surveys and studies of labels on processed, packaged food. In 2010 FDA reported the results of its most recent Food Label and Package Survey (FLAPS) of claims used on food packages in grocery stores nationwide. The most recent FLAPS, conducted during 2006 and 2007, was the 13th in a series of labeling studies and the first to include qualified health claims.

FDA's Office of Regulatory Affairs carries out food safety and labeling compliance inspections and enforcement activities. When the office identifies a labeling violation, FDA may send a warning letter—a notice that an enforcement action may be forthcoming if corrections are not made—or, for a less serious violation, an untitled letter communicating that corrective action is needed. At any point during the oversight process, FDA may hold a regulatory meeting with the company to resolve the labeling violation. For a serious labeling violation, FDA may ask a company to voluntarily recall food that has already entered the distribution chain.[5] When a violation is not corrected, FDA may initiate actions to seize and remove the food from the marketplace (seizure) or enjoin a company from continuing the practice that violates food labeling statutes and regulations (an injunction). For food imported from a foreign country, FDA may issue an import refusal to prevent a shipment with a serious labeling violation from entering the United States. FDA may also release a shipment "with comment"—that is, allow the shipment with a labeling violation to enter the United States with a notice to the importer that subsequent shipments might be refused entry if the violation is not corrected. In addition, FDA maintains "import alerts" that flag entries of imported foods that appear to be violative and may be detained, based on, for example, significant recurring or unusual violations.

Since January 2007, federal oversight of food safety has been on GAO's high-risk list of government programs that need broad-based transformation to achieve greater economy, efficiency, effectiveness, accountability, and sustainability.[6] Our September 2008 report on food labeling discusses in detail FDA's oversight of food labeling laws and regulations.[7] As we reported, FDA had little assurance that companies complied with food labeling laws and regulations for preventing false or misleading labeling, among other things. We found weaknesses in FDA's oversight and use of data and resources. For example, although FDA's inspection guidance directs inspectors to examine three product labels during a food facility inspection, FDA did not have reliable data on the number of labels examined or which inspections included label examinations. In following up on our recommendations in that report, we found that FDA has taken action on four of the report's seven recommendations and plans to continue addressing our remaining recommendations for better leveraging its resources and staff as resources become available.

# FDA HAS ALLOWED THE USE OF 12 QUALIFIED HEALTH CLAIMS ON FOOD LABELS AND PROVIDED LIMITED OVERSIGHT FOR THESE CLAIMS

Since 2002, when it announced its decision to allow companies to petition the agency for qualified health claims on food, FDA has determined that it would allow the use of 12 qualified health claims for food, and spent at least an estimated $12.8 million since 2000 on health claim and qualified health claim activities. FDA data indicate that companies' interest in qualified health claims has slowed. In addition, FDA has exercised limited oversight of industry's use of these claims.

## FDA Has Allowed the Use of 12 Qualified Health Claims for Food

From 2002 through September 2010, FDA received 16 petitions proposing 60 claims, which it considered for qualified health claims on food labels.[8] After reviewing the scientific evidence presented in the petitions, FDA determined that there was enough credible scientific evidence for the agency to modify 12 health claims with qualifying language to characterize in detail the strength and limitations of the scientific support for those claims. In letters to petitioners, FDA stated that it intended to consider exercising its enforcement discretion to *not* enforce certain requirements of the Federal Food, Drug, and Cosmetic Act and its implementing regulations for these qualified health claims. The claims had to meet the other requirements for health claims and any claim-specific conditions stipulated by FDA. FDA posted the 12 qualified health claims and conditions for their use on its Web site. Table 1 lists the 12 qualified health claims that are available for use on food labels.

In addition, FDA denied 42 of the 60 proposed claims. In letters to the petitioners conveying these decisions, FDA discussed its consideration of the specific scientific evidence submitted to support the claims it denied and how it reached a final determination that "neither a disclaimer nor qualifying language would suffice to prevent consumer deception" if the claims were used. FDA attributed each of the denials to the lack of credible scientific evidence to support the claim. In addition to the 42 claims that were denied and the 15 used to develop the 12 allowed qualified health claims, 1 proposed claim is pending, and 2 were withdrawn.

**Table 1. Twelve Qualified Health Claims Allowed for Use on Food, through September 2010**

| Food/disease risk | Qualified health claim |
|---|---|
| Tomatoes/reduced risk for prostate cancer | Very limited and preliminary scientific research suggests that eating one-half to one cup of tomatoes and/or tomato sauce a week may reduce the risk of prostate cancer. FDA concludes that there is little scientific evidence supporting this claim. |
| Tomatoes/reduced risk for ovarian cancer | One study suggests that consumption of tomato sauce two times per week may reduce the risk of ovarian cancer, while this same study shows that consumption of tomatoes or tomato juice had no effect on ovarian cancer risk. FDA concludes that it is highly uncertain that tomato sauce reduces the risk of ovarian cancer. |
| Tomatoes/reduced risk for gastric cancer | Four studies did not show that tomato intake reduces the risk of gastric cancer, but three studies suggest that tomato intake may reduce this risk. Based on these studies, FDA concludes that it is unlikely that tomatoes reduce the risk of gastric cancer. |
| Tomatoes/reduced risk for pancreatic cancer | One study suggests that consuming tomatoes does not reduce the risk of pancreatic cancer, but one weaker, more limited study suggests that consuming tomatoes may reduce this risk. Based on these studies, FDA concludes that it is highly unlikely that tomatoes reduce the risk of pancreatic cancer. |
| Green tea/reduced risk for breast cancer | Two studies do not show that drinking green tea reduces the risk of breast cancer in women, but one weaker, more limited study suggests that drinking green tea may reduce this risk. Based on these studies, FDA concludes that it is highly unlikely that green tea reduces the risk of breast cancer. |

| Food/disease risk | Qualified health claim |
|---|---|
| Green tea/reduced risk for prostate cancer | One weak and limited study does not show that drinking green tea reduces the risk of prostate cancer, but another weak and limited study suggests that drinking green tea may reduce this risk. Based on these studies, FDA concludes that it is highly unlikely that green tea reduces the risk of prostate cancer. |
| Nuts/reduced risk for heart disease | Scientific evidence suggests but does not prove that eating 1.5 ounces per day of most nuts [such as *name of specific nut*] as part of a diet low in saturated fat and cholesterol may reduce the risk of heart disease. [See nutrition information for fat content.] |
| Walnuts/reduced risk for coronary heart disease | Supportive but not conclusive research shows that eating 1.5 ounces per day of walnuts, as part of a low saturated fat and low cholesterol diet and not resulting in increased caloric intake may reduce the risk of coronary heart disease. See nutrition information for fat [and calorie] content. |
| Omega-3 fatty acids/reduced risk for coronary heart disease | Supportive but not conclusive research shows that consumption of EPA and DHA omega- 3 fatty acids may reduce the risk of coronary heart disease. One serving of [Name of the food] provides [ ] gram of EPA and DHA omega-3 fatty acids. [See nutrition information for total fat, saturated fat, and cholesterol content.][a] |
| Monounsaturated fatty acids from olive oil/reduced risk for coronary heart disease | Limited and not conclusive scientific evidence suggests that eating about 2 tablespoons (23 grams) of olive oil daily may reduce the risk of coronary heart disease due to the monounsaturated fat in olive oil. To achieve this possible benefit, olive oil is to replace a similar amount of saturated fat and not increase the total number of calories you eat in a day. One serving of this product contains [x] grams of olive oil. |
| Unsaturated fatty acids from canola oil/reduced risk for coronary heart disease | Limited and not conclusive scientific evidence suggests that eating about 1½ tablespoons (19 grams) of canola oil daily may reduce the risk of coronary heart disease due to the unsaturated fat content in canola oil. To achieve this possible benefit, canola oil is to replace a similar amount of saturated fat and not increase the total number of calories you eat in a day. One serving of this product contains [x] grams of canola oil. |
| Corn oil/reduced risk for heart disease | Very limited and preliminary scientific evidence suggests that eating about 1 tablespoon (16 grams) of corn oil daily may reduce the risk of heart disease due to the unsaturated fat content in corn oil. FDA concludes that there is little scientific evidence supporting this claim. To achieve this possible benefit, corn oil is to replace a similar amount of saturated fat and not increase the total number of calories you eat in a day. One serving of this product contains [x] grams of corn oil. |

Source: FDA.

Note: The 12 claims are based on 15 proposed claims.

[a] The omega-3 fatty acids, EPA (eicosapentaenoic acid) and DHA (docosahexaenoic acid), are components of some fatty fish (primarily cold-water fish), fish oils, other foods (e.g., seaweed), and food ingredients (e.g., algal oils).

## FDA Officials Estimated Spending at Least $12.8 Million to Implement and Administer Health Claims and Qualified Health Claims since 2000

Since FDA decided to allow companies to petition the agency for qualified health claims, it has carried out a number of activities to implement the framework for qualifying claims and making scientific determinations on proposed claims. These activities included, among others, developing guidance for industry and FDA staff; carrying out research; and analyzing studies, surveys, and other scientific evidence submitted in support of the proposed claims. FDA officials estimated that, for the 11 fiscal years 2000 through 2010, CFSAN has spent at least $12.8 million, including expenses for nearly 88.7 full-time equivalent (FTE) staff, to implement and administer health claims and qualified health claims. FDA officials provided estimates because they said that the agency does not maintain data on expenditures and staff charges for the different responsibilities performed by CFSAN's Office of Nutrition, Labeling, and Dietary Supplements staff. However, for offices outside the Office of Nutrition, Labeling, and Dietary Supplements that worked on qualified health claim activities for food labeling, such as the Office of the Chief Counsel, FDA officials could not provide estimates with any degree of confidence on expenditures and staff years spent on activities related to health or qualified health claims since 2000. It is also unclear what portion of the estimated $12.8 million was used solely for qualified health claim activities and what portion was used for regulatory activities on the 12 health claims for food. However, in FDA's technical comments on a draft of this report, the agency pointed out that many of the health claims assessed under the significant scientific agreement standard were completed prior to 2000.[9]

## Companies Minimally Use Qualified Health Claims

FDA data and other research indicate that companies' interest in qualified health claims has slowed. FDA's and industry's initial expectation of a flood of petitions for qualified health claims on food was never realized, according to agency officials. As shown in table 2, three-quarters (45 of 60) of the proposed claims for food labels were submitted in five petitions in 2004, and only one petition with one proposed claim was submitted after 2006.

**Table 2. Petitions and Proposed Qualified Health Claims for Food Labeling Submitted Annually, Calendar Year 2002 through September 30, 2010**

| Calendar year | Number of petitions submitted | Number of proposed claims submitted |
|---|---|---|
| 2002 | 2 | 3 |
| 2003 | 4 | 4 |
| 2004 | 5 | 45 |
| 2005 | 2 | 2 |
| 2006 | 2 | 5 |
| 2007 | 0 | 0 |
| 2008 | 0 | 0 |
| 2009 | 1 | 1 |
| 2010 | 0 | 0 |
| **Total** | **16** | **60** |

Source: GAO analysis of FDA data.

In addition, research by FDA and others shows that companies are making minimal use of qualified health claims. In 2010, FDA reported its latest FLAPS results of claims on food labels in grocery stores.[10] As shown in table 3, few food labels—only 0.4 percent—had qualified health claims.

**Table 3. Types of Claims FDA Found on Food Labels in Its Survey of Grocery Stores**

| Claim type | Percentage of labels with claim type |
|---|---|
| Qualified health claim | 0.4 |
| Health claim (based on significant scientific agreement or an authoritative source) | 4.3 |
| Structure/function claim | 5.5 |
| Other implied nutrient content claim[a] | 5.6 |
| "Healthy" claim (an implied nutrient content claim)[a] | 7.6 |
| Significant source claim[a] | 20.5 |
| Nutrient content claim | 53.2 |

Source: FDA survey conducted in 2006-2007.

Note: A sampled label may have had multiple claims. The claim types are those used by FDA in the survey.

[a] A type of nutrient content claim.

Moreover, according to a 2009 study by West Virginia University researchers, companies that could have used qualified health claims more often did not.[11] This study focused on labels of foods *eligible* to use qualified health claims. Specifically, it found that only 4.6 percent of food packages *eligible* for a qualified health claim actually used one. Instead, companies chose to use a different type of claim to convey the health or nutritional benefits of their food.

## FDA Oversight of Qualified Health Claims Is Limited

FDA continues to use its enforcement discretion for qualified health claims and has not taken enforcement actions against companies for violations involving these claims. However, as part of an oversight initiative targeting food labeling compliance, FDA issued 17 warning letters to food companies, from December 2009 to February 2010, including 2 for violations related to their use of qualified health claims. FDA officials stated that to the best of their knowledge these were the first warning letters the agency had issued regarding qualified health claims on food.

The two warning letters for qualified health claims were issued to Fleminger, Inc., for claims about green tea, and Diamond Food, Inc., for claims about walnuts. As shown in table 4, according to FDA's warning letters, the companies (1) failed to use the specific qualified health claims statements allowed by FDA and (2) added language that was not supported by scientific evidence. Specifically, according to FDA's letter to Fleminger, Inc., the company's Web site had stated that high consumption of green tea is associated with reduced risks for cancer of the esophagus, stomach, colon, rectum, pancreas, urinary bladder, lung, liver and ovary— claims regarding diseases that FDA did not find support for, and did not include in the qualified health claims. Noting that the statements on the Web site altered the meaning of the qualified health claim language and misrepresented FDA's conclusions, FDA found that the statements caused the products to be misbranded in violation of the Federal Food, Drug, and Cosmetic Act. In its warning letter to Diamond Food, Inc., FDA said, among other things, that the product label contained health claims that were not authorized by FDA, in violation of the Federal Food, Drug, and Cosmetic Act. The label had the phrase "OMEGA-3 2.5 g per serving" and heart symbols adjacent to information about the amount of omega-3 in the product—implying that omega-3 reduces the risk of coronary heart disease and that the relationship between walnuts and coronary heart disease is related to the omega-3 fatty acid content of walnuts. However, as FDA's letter points out, there is not sufficient evidence to identify omega-3 fatty acid as the substance in walnuts that reduces the risk of coronary heart disease.

**Table 4. Two FDA Warning Letters Issued in February 2010 Regarding Qualified Health Claims**

| Company, date of warning letter, and product cited for violation | Qualified health claims allowed for product | Excerpts from the warning letter |
|---|---|---|
| Fleminger, Inc. (Feb. 22, 2010) TeaForHealth™ green tea products Dr. Lee's TeaForHealth® 710EGCG™ inabottle™ Green Tea and TeaForHealth® 710EGCG™ Ready-To- Drink Natural Brewed Green Tea | 1. "Two studies do not show that drinking green tea reduces the risk of breast cancer in women, but one weaker, more limited study suggests that drinking green tea may reduce this risk. Based on these studies, FDA concludes that it is highly unlikely that green tea reduces the risk of breast cancer." 2. "One weak and limited study does not show that drinking green tea reduces the risk of prostate cancer, but another weak and limited study suggests that drinking green tea may reduce this risk. Based on these studies, FDA concludes that it is highly unlikely that green tea reduces the risk of prostate cancer." | The letter states that the following claims, presented on the company's Web sites, were unauthorized and not consistent with either of the qualified health claims FDA allows for green tea: Examples of health claim on www.tea forhealth.com include the following: "Green tea may reduce the risk of breast and prostate cancers. FDA has concluded that there is credible evidence supporting this claim although the evidence is limited." Examples of health claim on www.greentea-haus. com in the form of headings categorized as "educational materials" include the following: "Epidemiological and clinical studies on the relationship between cancer risk and the consumption of green tea . . ." Examples of health claims in "The Truth in Tea" include the following:<br>• "[H]igh consumption of green tea [is] associated with reduced cancer rates of the breast, esophagus, stomach, colon, rectum, pancreas, urinary bladder, prostate, lung, liver and ovary . . ."<br>• "Recent medical research has provided evidence that drinking green tea may reduce the risk of fatal heart attack, stroke, Alzheimer's disease, Parkinson's disease, help reduce body fat and help fight viral infection." |

**Table 4. (Continued)**

| Company, date of warning letter, and product cited for violation | Qualified health claims allowed for product | Excerpts from the warning letter |
|---|---|---|
| Diamond Food, Inc. (Feb. 22, 2010) Diamond of California Shelled Walnuts | "Supportive but not conclusive research shows that eating 1.5 oz of walnuts per day, as part of a low saturated fat and low cholesterol diet, and not resulting in increased caloric intake, may reduce the risk of coronary heart disease. Please refer to nutrition information for fat content and other details about the nutritional profile of walnuts." | The letter states that the following statement on the product label is an unauthorized health claim: "The omega-3 in walnuts can help you get the proper balance of fatty acids your body needs for promoting and maintaining heart health. In fact, according to FDA, supportive but not conclusive research shows that eating 1.5 oz of walnuts per day, as part of a low saturated fat and low cholesterol diet, and not resulting in increased caloric intake, may reduce the risk of coronary heart disease" The letter further states that the statement suggests that the evidence supporting a relationship between walnuts and coronary heart disease is related to the omega-3 fatty acid content of walnuts. There is not sufficient evidence to identify a biologically active substance in walnuts that reduces the risk of coronary heart disease. |

Source: FDA documents.

Note: FDA's warning letters to both companies also cited violations other than those for qualified health claims; these violations are not shown in the table.

In a March 2010 open letter to industry, the FDA Commissioner noted that the warning letters did not attempt to cover all products with labeling violations, but that they did cover a range of concerns about how false or misleading labels can undermine the "intention of Congress to provide consumers with labeling information that enables consumers to make informed and healthy food choices." The warning letters gave companies 15 working days from the receipt of the letters to respond with the steps that they are or will be taking to correct their label violations. According to FDA officials, all the companies responded to the warning letters. When FDA has verified that the violations have been addressed, it will issue, and post to its Web site, closeout letters indicating that corrective actions were taken, according to the officials. As of December 23, 2010, FDA had not issued closeout letters for any of the 17 companies that received these warning letters for labeling violations. (App. V summarizes the other 15 warning letters and an untitled letter that FDA issued regarding claims found on food labels during the December 2009 through February 2010 oversight initiative.)

## Consumers Have Difficulty Understanding the Level of Scientific Support for Qualified Health Claims and How These Claims Differ from other Claims

Research has shown, and stakeholders have indicated, that consumers find it difficult to understand the differences between qualified health claims and health claims with significant scientific agreement. Consumers have similar difficulties understanding the differences among health, structure/function, and other health- and nutrient-related claims. European Union (EU) and Canadian food labeling officials, noting the potential for qualified health claims to confuse or mislead consumers about the benefits of a food, stated that their governments do not allow the use of qualified health claims on food labels.

### Consumers Have Difficulty Understanding the Level of Scientific Support for Qualified Health Claims, According to Research and Stakeholders

Research conducted over the past decade by FDA, FTC, and the International Food Information Council on consumers' understanding of qualified health claims consistently shows that consumers find it difficult to understand the degree of scientific support for qualified health claims on food labels.[12] In particular, according to our analysis of this research, consumers cannot consistently distinguish between health claims, which have significant scientific agreement, and qualified health claims, which have lower levels of scientific evidence (i.e., good/moderate and low levels of scientific support). For example,

- According to a 2009 FDA study, which was based on a Web-based survey of over 7,000 adults, the respondents could not distinguish between the strength of scientific evidence for claims with significant scientific agreement and for claims with good/moderate levels of support.[13] Specifically, respondents' ratings of scientific support for two health claims with substantial scientific support (i.e., the benefits of soluble fiber and of stanol esters from vegetable oil in reducing the risk of heart disease) were not consistently different than ratings for the qualified health claims for omega-3 fatty acids and heart disease, which have good/moderate scientific support.[14]

- FDA's 2009 study also found that respondents could not consistently distinguish between the level of support for qualified claims related to (1) omega-3 fatty acids for reducing the risk of heart disease, which has good/moderate scientific support, and (2) monosaturated fat from olive oil for reducing the risk of heart disease, which has a low level of scientific support.
- The International Food Information Council's 2008 research on consumers' understanding of qualified health claims, based on a Web-based survey of 5,642 adults, showed that words such as "inconclusive" meant different things to different consumers.[15] According to the council, some consumers believed that "inconclusive" implied "honesty" and "believability," while others believed that it implied extreme negativity. However, the phrase "not conclusive" is used in four qualified health claims, including the walnut claim: "Supportive but not conclusive research shows that eating 1.5 ounces per day of walnuts, as part of a low saturated fat and low cholesterol diet and not resulting in increased caloric intake, may reduce the risk of coronary heart disease."
- The council's research also showed that consumers had trouble distinguishing the four distinct levels of scientific support behind claims— that is, distinguishing among the (1) significant scientific agreement for health claims and, for qualified health claims, the (2) moderate scientific evidence, (3) limited or low scientific evidence, and (4) little or extremely low scientific evidence. Regardless of the different language options used, 78 percent of consumers ranked the four levels incorrectly.
- FTC's 2005 study presented the findings of research on consumers' perceptions of print advertisements containing qualified health claims, obtained from interviews of 480 adults, in eight geographically dispersed shopping mall facilities.[16] According to FTC officials, although it might be possible to communicate the proper order of

various levels of qualification using strongly worded language or graphical displays, it would nonetheless be difficult to develop qualifying language in advertising to communicate a low level of scientific certainty. The survey results indicated that none of the tested language, whether appearing in real or fictitious product advertisements, communicated serious limitations in scientific evidence. In addition, consumers interpreted all of the tested advertisements in a disparate fashion. For example, in an advertisement for a fictitious antioxidant vitamin supplement, about two-thirds of the consumers responding either overestimated or underestimated the certainty of the science.

Furthermore, on the basis of its own research as well as FDA's, FTC has continued to express concern about the potential for qualified health claims to mislead consumers. In 2009, the Director of FTC's Bureau of Consumer Protection announced that one of the bureau's primary goals was to prevent claims made in food advertising that were not supported by the weight of the scientific evidence, noting that one outlier study should not be the sole basis of a claim, particularly for a health benefit. [17] According to FTC's Enforcement Policy Statement of Food Advertising, the commission believes that qualified claims based on evidence that is inconsistent with the larger body of evidence have the potential to mislead consumers and therefore are likely to violate the Federal Trade Commission Act. Further noting that "objective claims carry with them the implication that they are supported by valid evidence," in its policy statement, FTC recognized that qualified health claims may be permitted in certain limited instances—if they expressly convey "clearly and fully" the extent of scientific support. However, as noted in the statement, if such claims are *not* consistent with the majority of evidence, they could potentially mislead consumers and, under its legal framework, FTC would find them deceptive.

If consumers cannot properly distinguish among the claims, they may be unable to make informed decisions about the likelihood that a food will have the claimed health benefit. Some consumer and health association stakeholders told us that qualified health claims should be eliminated because of consumers' confusion, although a few suggested that the claims with moderate scientific support should remain. In addition, in a January 2007 letter to FDA, the American Medical Association stated that it continues to vigorously oppose the use of qualified health claims on foods; that FDA has no basis for allowing them because, according to the association, *Pearson v. Shalala* does not apply to foods; and that recent research has shown that qualifying language cannot remedy the possible deceptiveness of qualified health claims. On the other hand, the Grocery Manufacturers Association supports the continued use of qualified health claims, and according to the association representatives, some members find them to be a useful advertising tool.

In part because of the results of FDA's research on consumers' understanding of qualified health claims, the European Union does not allow these claims on food labels, according to EU labeling officials. These officials said that these claims may confuse or mislead consumers. Similarly, Canada does not allow these claims because of their potential to confuse or mislead consumers, according to a Canadian food labeling official.

## Research and Stakeholders Find Consumers Have Difficulty Distinguishing among Health, Qualified Health, Structure/ Function, and Nutrient Content Claims

According to research conducted by FDA, the International Food Information Council, and academia, consumers have difficulty distinguishing among the many different types of claims on food labels, including health claims, qualified health claims, structure/function claims, and nutrient content claims. For example,

- According to a 2008 article on how 1,077 participants in a 2006 FDA study interpreted health messages on food labels, when respondents were well acquainted with the nutrient or diet-disease relationship—such as the link between calcium and osteoporosis—there was no difference in how strongly respondents believed in the stated health benefit, regardless of whether the claim was a health, structure/function, or nutrient content claim.[18]
- According to the International Food Information Council's 2008 study, consumers rate the level of scientific evidence and other attributes associated with a product containing a structure/function claim as similar to the evidence and other attributes of health claims with significant scientific agreement on a product. Specifically, the council found that consumers are just as likely to purchase a product with a structure/function claim, which FDA does not review, as they are to purchase a product with a health claim supported by significant scientific agreement, which FDA does review. Furthermore,

structure/function claims were perhaps the most popular of all the claims the council tested: Most consumers liked their brevity and general health messages more than health claims, which they saw as too wordy and too disease specific. For example, several consumers mentioned that they did not like to see the word "cancer" on the labels for their food products, even if the label asserted the food would help prevent cancer.

- Academic researchers also noted that consumers tend to interpret the lack of disclaimers (statements that FDA has evaluated the claims on certain dietary supplements) as meaning that the claim had received FDA's review and approval.[19]

FDA reviews some health- and nutrient-related claims on food, but many are not reviewed (e.g., structure/function claims), and as a result, public health may not be adequately protected, according to a 2010 report by the National Academies' Institute of Medicine (IOM). According to this IOM report, consumers have difficulty assessing the scientific merits of these claims. Similarly, some health and consumer stakeholders we interviewed said that FDA should particularly review structure/function claims because consumers do not understand the difference between them and health claims that do receive FDA's review. They also said that like health claims, structure/ function claims should have scientific standards to ensure they are not false or misleading and that these claims should not be allowed on foods that are high in sugar, saturated fat, or sodium.

In addition, according to FDA officials, the agency would be better able to take enforcement action against companies that make false or misleading structure/function claims if it had express authority to access company research. Furthermore, a contributing author of the IOM report explained that while FDA can conduct its own studies as resources permit, it needs authority to require food companies to provide studies focused on specific claims and products. Some stakeholders, such as representatives from the American Society for Nutrition and the Center for Science in the Public Interest, said that FDA should have authority to require consumer and scientific studies from industry so that industry, rather than FDA, bears the burden of proof that a claim is truthful and not misleading. However, a representative from the American Dietetic Association expressed concern that giving FDA access to companies' research could hinder industries' willingness to conduct research, while a representative from the Grocery Manufacturers Association opposed any suggestions that FDA be given additional authority to require proprietary studies from its members.

# INDUSTRY MORE WIDELY USES STRUCTURE/FUNCTION CLAIMS ON FOOD WITH MINIMAL FDA OVERSIGHT, BUT FTC HAS TAKEN SOME ACTION WHEN THESE CLAIMS WERE DECEPTIVE

While FDA data indicate that industry's interest in qualified health claims on food labels has slowed to one petition for such a claim since 2006, data also indicate that industry more widely uses structure/function claims to communicate health benefits to consumers, with minimal FDA oversight. However, FTC has taken action to cause a major food company to withdraw deceptive structure/function claims on its labels and advertising.

## Industry More Often Uses Structure/Function Claims on Food with Minimal FDA Oversight

As noted earlier and in table 3, FDA's FLAPS results found only about 0.4 percent of food labels in grocery stores had qualified health claims. In contrast, about 4.3 percent had health claims, and 5.5 percent had structure/function claims. The most common claims found in the FLAPS study were nutrient content claims or variations of these claims. Consumers can usually check the nutrition facts panels or ingredients lists, which are required on food labels, for corroborating information for nutrient content claims. Although consumers are able to verify information from nutrient facts panels, they do not have such information readily available on labels for health claims, qualified health claims, or structure/function claims.

Also, the 2009 study by West Virginia University researchers showed that companies that could have used qualified health claims chose more than twice as often to use structure/function claims.[20] This study focused on labels of foods *eligible* to use qualified health claims. Specifically, it found that only 4.6 percent of food packages *eligible* for a qualified health claim actually used one, while 9.4 percent of these *eligible* packages used structure/function claims instead. The study identified possible reasons for choosing structure/function claims more often than qualified health claims; for example,

(1) structure/function claims, unlike qualified health claims, do not mention diseases; (2) companies do not want consumers to relate unpleasant or unfavorable associations (such as cancer or colon polyps) with their products; and (3) the qualifying language FDA used in the qualified health claims is more limiting than companies had hoped. In addition, research funded by the food industry has shown that consumers prefer to see shorter, simpler claims on food labels. Also, companies are not required to submit scientific support or obtain FDA's premarket review and approval before using a structure/function claim on a food label.

As we said earlier, FDA is responsible for ensuring the proper labeling of food, and the use of structure/function claims is subject to the general statutory requirement that labeling not be false or misleading. However, FDA has not given companies guidance on the scientific support needed to prevent false or misleading information in a structure/function claim for food. FDA has guidance for the dietary supplements industry on the scientific support needed to prevent false or misleading information for a structure/function claim for dietary supplements.[21] In technical comments, FTC staff stated that "FDA could issue a statement that the same principles apply to foods."

One way for FDA to identify potentially false or misleading structure/function claims would be by including an examination of these claims in food facility compliance inspections. Following the *Compliance Program Guidance Manual*, inspectors are instructed to examine any three food labels from a facility. Inspectors are to focus their label reviews on violations, such as failure to declare allergens and failure to provide nutrition information. For nutrient content claims on food labels, inspectors can check the facts panels and ingredients lists to help corroborate the claims. Their inspection guidance manual identifies the statutes and regulations for health claims and nutrient content claims but does not provide instructions for identifying potentially false or misleading structure/function claims for food.

If FDA determines that a structure/function claim is false or misleading, it may send a warning letter to the company. Although FDA officials could not identify any warning letters or enforcement actions for such violations on food labels, they provided two warning letters issued in 2001 for nutrient content claim violations, in which FDA also expressed concerns with structure/function claims made by the companies on their food labels.[22] Specifically, those warning letters mentioned the following structure/function claim concerns:

- One warning letter, for New Morning Organic Ginseng Crunch and Organic GinkgOs, noted the label claims “for mental concentration, physical vitality and energy and its . . . anti-oxidant qualities” and “to sustain memory.” According to FDA's warning letter, the foods were represented as cereals. The letter noted that the claimed effect must be achieved through the nutritive value of a food, and that otherwise the product would be considered a drug under the Federal Food, Drug, and Cosmetic Act, and therefore subject to FDA’s drug regulations.
- The second warning letter, for Hansen’s Healthy Start Immune Juice, had a label claim stating that “Echinacea . . . may help stimulate the body’s production of interferon.” FDA again stated that the claimed effect had to be achieved through the nutritive value of the food and that otherwise the product would be considered a drug under the Federal Food, Drug, and Cosmetic Act, and therefore subject to FDA’s drug regulations.

FDA officials told us that these warning letters were issued about 16 months after an FDA-sponsored stakeholder meeting in 2000, which was convened to seek input for what FDA viewed at the time as a need for immediate attention to the growing use of structure/function claims and the serious oversight dilemma they posed for FDA. At the meeting, the stakeholders—from FDA, industry, consumer groups, and academia—identified an urgent need for regulating and overseeing structure/function claims and noted that FDA could not wait for legislative changes. The stakeholders stated, among other things, that

- consumers cannot be confident that the claims are truthful and not misleading because their scientific support is unclear,
- some consumers lack the tools or understanding needed to evaluate claim messages and understand what they mean, and
- consumers cannot clearly distinguish among health claims, structure/function claims, and other claims.

In a January 2001 “Letter to Manufacturers,” FDA reminded food companies that, among other things, “the claimed structure/function effects for foods must be achieved through nutritive value. If [they are] not, the product is subject to regulation as a drug.”

Subsequently, in a 2006 *Federal Register* notice, FDA requested comments on a series of questions on, among other things, whether it should require companies to notify FDA within 30 days before marketing a food with a structure/function claim, and whether FDA should require a disclaimer stating that FDA had not reviewed or approved the claim. As we noted earlier, federal law requires such notification and a disclaimer on dietary supplements for all structure/function claims that reads, "This statement has not been evaluated by the Food and Drug Administration. This product is not intended to diagnose, treat, cure, or prevent any disease." In commenting on these issues, two consumer groups supported requiring companies to submit structure/function claims for foods with novel ingredients (e.g., ingredients that are added to and not normally found in the food) to FDA for premarket approval. One of the consumer groups noted that disclaimers for dietary supplements have been shown to be ineffective, and thus the group was not suggesting they be used on food. A few industry groups commented that they believed FDA lacked authority to require a disclaimer or premarket approval. FDA has not adopted a disclaimer or notification requirement for structure/function claims on food labels.

Unlike FTC, which may require companies to provide the evidence to support their advertising claims, FDA bears the burden of proving that a structure/function claim is false or misleading without having the authority to compel companies at the investigation stage to produce the evidence that the companies assert as support for their advertising claims. According to senior attorneys in FDA's Office of the Chief Counsel, FDA's efforts to meet that burden are hampered by the lack of access to the evidence that a company relies on to make such a claim. In particular, while FDA may ask a company to provide its evidence, FDA does not have express legal authority to compel the company to provide information. To support an action, FDA may need to review the scientific literature to determine whether the literature supports a claim—an effort that may be time consuming and labor intensive. FDA may also need to establish through research how consumers perceive the claim. If the structure/function claim, or any claim for that matter, is one for which there is little publicly available information, it can be difficult for FDA to develop the evidence needed to support an enforcement action. It also can be difficult to evaluate the support for a structure/function claim when the functions, such as "immunity" or "attention," do not have generally established measures for the claimed benefit. For example, it is not necessarily clear what outcome measures would be a valid means of assessing whether a food "improves vitality."

Furthermore, in contrast to the United States, the EU follows a different course: EU Commission officials told us the responsible EU food agency must preapprove the type of food claims that FDA calls structure/function claims before they can be used. EU Commission officials also stated that their scientific review agency rejected 85 percent of structure/function claims companies have submitted because of the lack of scientific support. Canada is also reviewing structure/function claims food companies have submitted by developing a list of approved claims; it too has rejected most claims, including some found on labels in the United States, according to a Canadian official. To protect the public from false or misleading claims, the Canadian food labeling agency is working with the Canadian agency that regulates advertising to develop a framework for regulating structure/function claims.

## FTC Has Taken Some Enforcement Action on Deceptive Structure/ Function Claims on Food

Although FTC, operating under its memorandum of understanding with FDA, has not historically focused on claims on food labels, it has recently emphasized that health claims on food must be adequately substantiated and presented in a manner that is truthful and not deceptive, according to FTC officials. To clarify how its authority related to FDA's labeling regulatory scheme under NLEA, FTC issued an Enforcement Policy Statement on Food Advertising in May 1994 that provided the legal framework for its related enforcement actions. Under that framework, FTC would find an advertisement deceptive and therefore unlawful, if

- the advertisement contained a representation or omission of a material fact that is likely to mislead consumers acting reasonably under the circumstances, or
- the advertiser did not possess and rely on reasonable substantiating evidence of the claim's truth.

FTC has filed complaints on claims made in food advertising that it alleged to violate the Federal Trade Commission Act, and taken enforcement actions for alleged false structure/function claims on food labels and in advertisements.[23] Specifically, in 2009, FTC reached a settlement with Kellogg on a complaint alleging that the company made unsubstantiated claims that Frosted Mini-Wheats was "clinically shown to improve kids' attentiveness by nearly 20 percent." The company agreed to refrain from making comparable claims. In 2010, FTC reopened the complaint based on claims made by Kellogg that Rice Krispies cereal "now helps support your child's immunity." To resolve the investigation into the new claims, Kellogg agreed to new advertising restrictions and a revised settlement order.

FTC officials told us the commission could not have taken the actions that it did without authorities, such as access to companies' proprietary market and scientific research. They said FTC needed research and documentation from Kellogg to determine whether the information supported an enforcement action. FTC has authority to compel companies to provide evidentiary documents, among other things, and therefore it can get such documents on a voluntary basis. FTC obtained Kellogg's documents showing that the attentiveness claim was allegedly false. According to the FTC complaint, Kellogg's research showed that the cereal increased children's attentiveness by only about 11 percent.

FTC officials told us that they had informed FDA about their actions concerning Kellogg. However, because Kellogg cooperated, the two agencies did not have to coordinate further. Since September 2009, through monthly working group meetings, FDA and FTC have been coordinating to, among other things, prevent duplication of effort and share information on cases of mutual interest.

## CONCLUSION

Initially in 2002, FDA thought it would receive a flood of petitions for qualified health claims from food companies. However, this demand never materialized, and in fact, FDA has received only 16 petitions with 60 proposed claims and only 1 petition since 2006. Furthermore, the 12 qualified health claims that FDA has allowed are being minimally used by food companies. Structure/function claims are another matter, making up more than 5 percent of all food labels. FDA recognized a decade ago the growing use of such claims and the serious oversight dilemma they pose. Unlike FTC, which can require companies to provide the evidence to support their food advertising claims, FDA bears the burden of proving that a structure/function or other type of claim is false or misleading. Our work indicates that FDA's efforts to meet that burden are hampered by the lack of access to the scientific support that a company relies on to make such a claim. In particular, while FDA may ask a company to provide its support, FDA does not have express legal authority to compel the company to provide it. FTC, on the other hand, has authority to compel companies to provide evidentiary documents.

Furthermore, FDA has not given companies guidance on the level of scientific evidence needed to prevent false or misleading information in a structure/function claim for food; by not providing such guidance, FDA may not be doing all it can to help ensure that food labels are free from false or misleading claims. Moreover, FDA has not given its inspectors instructions in the *Compliance Program Guidance Manual* for identifying potentially false or misleading information in structure/function claims when examining food labels as part of food facility compliance inspections; without such instructions, inspectors may be missing one way for FDA to identify potentially false or misleading structure/function claims.

## Recommendations for Executive Action

To ensure that the health-related claims on food labels are not false or misleading to consumers, we recommend that the Secretary of Health and Human Services direct the Commissioner of FDA to take the following three actions:

- identify and request from Congress the authorities needed to access evidence from food companies regarding potentially false or misleading structure/function or other claims on food that would allow the agency to establish whether there is scientific support for the claims;
- provide guidance to industry on the type and strength of scientific evidence needed to prevent false or misleading information in a structure/function claim; and
- amend the *Compliance Program Guidance Manual* instructions to FDA inspectors for reviewing food labels during inspections of food facilities, to include steps for identifying potentially false or misleading structure/function claims for further review.

## Agency Comments and Our Response

We provided a draft of this report to the Secretary of Health and Human Services and FTC's General Counsel for their review and comment. In written comments, the Department of Health and Human Services' (HHS) Assistant

Secretary for Legislation stated that FDA welcomed the report for calling attention to the extremely complex and challenging legal and regulatory environment under which FDA works to protect consumers from false and misleading label claims on foods, particularly qualified health claims and structure/function claims. HHS stated that FDA also welcomed the report for calling attention to the differences in the legal authorities between FDA and FTC that allow FTC to compel companies to provide evidence supporting their advertising claims. FTC staff provided technical comments that we incorporated in the report, as appropriate.

With regard to our recommendations, HHS indicated that FDA generally agreed with our first two recommendations but believes that our third recommendation is impractical. Specifically, regarding our first recommendation—to identify and request from Congress the authorities needed to access evidence from food companies regarding potentially false or misleading structure/function or other claims on food that would allow the agency to establish whether there is scientific support for the claims—HHS commented that FDA would work to determine better approaches to enhancing its oversight of structure/function claims, including whether additional statutory authorities are needed.

Regarding our second recommendation—to provide guidance to industry on the type and strength of scientific evidence needed to prevent false or misleading information in a structure/function claim—HHS noted FDA's support for providing guidance to industry on the type of evidence needed. FDA agreed to consider developing such guidance as agency priorities and resources permit. FDA has guidance for the dietary supplements industry on the scientific support needed to prevent false or misleading information for a structure/function claim for dietary supplements. As we noted earlier, in technical comments, FTC staff stated that "FDA could issue a statement that the same principles apply to foods."

Regarding our third recommendation—to amend the *Compliance Program Guidance Manual* instructions to FDA inspectors for reviewing food labels to include steps for identifying potentially false or misleading structure/function claims during inspections of food facilities—FDA agreed that field investigators, to whom the *Compliance Program Guidance Manual* is targeted, have significant exposure to food labels that may contain any one of a myriad of structure/function claims during their routine inspections of food producers. However, the agency sees significant challenges to providing specific steps to investigators to identify potentially false or misleading structure/function claims during inspections. FDA found the recommendation impractical, noting that judging whether a structure/function claim is false or misleading requires knowledge of the science that would support or dispute the claim. FDA believes it would be impossible to provide inspectors with the knowledge necessary to make such an assessment in the field. FDA further states that, while structure/function claims are not highlighted as an area of emphasis, the *Compliance Program Guidance Manual* includes instructions for inspectors to collect information to determine the extent to which food products are in compliance with all labeling requirements and provides additional references regarding the types of claims permitted on food labels. We agree that the guidance manual includes several instructions to help inspectors review food labeling. It contains, for example, the requirements for the nutrition facts panel and identifies allergens that must be declared. The guidance manual also identifies the statute and regulations for health claims and nutrient content claims, as well as the statute and regulations prohibiting food labeling that is false or misleading. However, the *Compliance Program Guidance Manual* does not provide instructions to help inspectors identify potentially false or misleading structure/function claims on food. While we agree it may be impractical for field inspectors to judge whether a claim is false or misleading, we believe that FDA can amend the instructions for inspectors with steps to help them identify potentially false or misleading structure/function claims that may warrant further FDA review—leaving the determinations of whether the claims are indeed potentially false or misleading to FDA staff with the necessary science backgrounds. We clarified our recommendation.

FDA also provided technical comments that we incorporated in the report, as appropriate.

Lisa Shames
Director, Natural Resources and Environment

## APPENDIX I: FDA'S ADMINISTRATION OF HEALTH CLAIMS IN RESPONSE TO *PEARSON V. SHAL ALA*

In 2008 we reported on the Food and Drug Administration's (FDA) efforts to help consumers choose healthy foods through the information provided on food labels.[24] In that report we presented information on FDA's administration of health claims, and specifically its treatment of qualified health claims, after the *Pearson v. Shalala* ruling. That information is repeated here with some modifications.

As we reported, the Nutrition Labeling and Education Act of 1990 (NLEA)[25] amended the Federal Food, Drug, and Cosmetic Act to include provisions that govern the use of health claims on food labeling.[26] For conventional foods, the NLEA requires that any claim that expressly or by implication characterizes the relationship of a nutrient to a disease or health-related condition must be authorized by the Secretary of Health and Human Services (delegated to FDA) through a regulation.[27] Under the NLEA, FDA may authorize a health claim for a conventional food if it determines, based on the totality of publicly available scientific evidence, that there is "significant scientific agreement" among experts—qualified by scientific training and experience to evaluate such claims—that the claim is supported by such evidence. Although the NLEA also provided for the use of health claims in dietary supplement labeling, Congress did not require health claims for dietary supplements to be subject to the same statutory procedures and standards as health claims for conventional food. Instead, health claims for dietary supplements were to be subject to procedures and standards established in regulations issued by the Secretary of Health and Human Services (delegated to FDA).

In 1991, FDA published a proposed rule in the *Federal Register*, proposing the implementation of statutory procedures and standards for health claims for conventional food, and proposing to adopt those same procedures and standards for dietary supplement health claims.[28]

However, before the rule could be finalized, Congress passed legislation that generally prohibited FDA from implementing the NLEA for dietary supplements until December 15, 1993.[29] Therefore, in January 1993, when FDA adopted the final rules for health claims for conventional foods, it did not finalize rules for health claims on dietary supplements.[30] However, 1 year later, after the prohibition of implementation of NLEA for dietary supplements had expired, FDA adopted a rule that subjected health claims for dietary supplements to the same general requirements that applied to conventional foods.[31] Under those rules, any person wanting to include a health claim on the label for a conventional food or dietary supplement must petition FDA for authorization before including the claim on the label. If FDA determines, based on the totality of publicly available information, that there is significant scientific agreement in support of that claim, it will authorize its use in regulation.

FDA's health claim regulations for dietary supplements were the subject of several lawsuits in the 1990s. In a case known as *Pearson v. Shalala*, the U.S. Court of Appeals for the District of Columbia Circuit held that the First Amendment does not permit FDA to prohibit a potentially misleading health claim on the label of a dietary supplement, unless FDA considers whether a disclaimer on the product's label could negate the potentially misleading nature of that claim.[32] Specifically, the court stated that although inherently or actually misleading information in food labeling or advertising may be prohibited, potentially misleading information cannot face an absolute prohibition. Instead, potentially misleading information may be regulated only if those regulations directly advance a substantial government interest and offer a reasonable fit between the government's goals and the means chosen to accomplish those goals. The court found a substantial interest in protecting the public health and preventing consumer fraud. However, it found that FDA's regulation requiring health claims to be supported by significant scientific agreement did not directly advance the interest in public health, and even though the regulations directly advanced the interest in preventing consumer fraud, the fit between the goals of the regulations and the means employed—an outright ban without the possibility of a disclaimer—was not reasonable.[33]

Following the decision in *Pearson v. Shalala*, FDA announced its plan to respond, stating that it would deny, without prejudice, all petitions for the use of health claims on dietary supplements that did not meet the significant scientific agreement standard while the agency conducted and completed a rulemaking to consider the procedures and standards governing such claims. Then, according to FDA, once a rule was finalized, the agency would revisit the petitions it had denied. However, in 2000, citing concerns over additional First Amendment challenges, FDA announced plans to modify that policy.[34] FDA stated that it would continue to approve health claims for dietary supplements that met the significant scientific agreement standard, but it would exercise its enforcement discretion and not take action against health claims for dietary supplements that failed to meet the standard under certain circumstances. Specifically, upon the submission of a valid petition for preapproval of a health claim for a dietary supplement, if FDA did not find significant scientific agreement, but, in evaluating the weight of the evidence, did find that the scientific evidence in support of the claim outweighed the scientific evidence against it, and consumer health and safety were not threatened, the agency would inform the petitioner of conditions under which the agency would refrain from taking enforcement action against the health claim. If the scientific evidence against the health claim outweighed the scientific evidence in support of it, FDA would deny any use of the health claim.

In 2002, the agency announced the availability of guidance, updating its approach to implementing the decision in *Pearson v. Shalala*.[35] In large part, the procedures remained the same; however, FDA included health claims for conventional foods under the procedures, even though *Pearson v. Shalala* directly addressed only dietary supplements. FDA stated that it believed that such a move would precipitate greater communication in food labeling and thereby enhance public health. In addition, FDA stated that including health claims for conventional foods in its enforcement discretion policy would help avoid further constitutional challenges. In 2003, FDA announced the availability of two new guidance documents describing interim procedures that, among other things, addressed a then recent U.S. District Court for the District of Columbia decision that found the weight of the evidence standard that FDA first articulated in guidance in 2000 was inappropriate.[36] According to the district court in that case, FDA should evaluate qualified health claims based on the presence of "credible evidence," not the weight of the evidence.[37]The 2003 guidance documents set forth new procedures for qualified health claims for conventional foods and dietary supplements. Specifically, qualified health claim petitions would be evaluated using an evidence-based ranking system that would rate the strength of the publicly available scientific evidence. A claim would be denied if there was no credible evidence to support it. Otherwise, based on the competent and reliable scientific evidence in support, a claim would be assigned to one of four ranked levels—the first level being "significant scientific agreement among qualified experts" and the remaining three levels being for claims supported by some lower level of credible evidence. Each of the three categories not ranked as supported by significant scientific agreement would correspond to one of three standardized qualifying statements (i.e., disclaimers).[38] So long as the qualified health claim bore the appropriate language, met other applicable health claim regulations, and adhered to criteria established in FDA's letter of enforcement discretion in response to the petition, FDA would exercise its enforcement discretion and refrain from acting against the health claim.

In November 2003, FDA published an Advance Notice of Proposed Rulemaking recognizing the need to establish transparent, long-term procedures that have the effect of law.[39] In that announcement, FDA presented several regulatory alternatives it might take, stating that it could (1) incorporate the interim procedures and evidence-based ranking system the agency described in its 2000 guidance into regulation; (2) subject health claims to notice-and-comment rulemaking, as before *Pearson v. Shalala,* but reinterpret the "significant scientific agreement" standard to refer to the evidence supporting the claim being made, instead of the underlying substance-disease relationship; or (3) treat qualified health claims as outside the NLEA and regulate them on a postmarket basis (i.e., pursue the product as misbranded if the health claim renders the label false or misleading because the claim lacks substantiation). FDA did not work on this proposed rulemaking in 2010.

FDA issued final guidance describing the evidence-based review system it intends to use to evaluate publicly available scientific evidence for health claims and qualified health claims in January 2009.[40] This guidance replaced the 2003 guidance document describing the evidence-based ranking system.[41] The 2009 guidance document set out an approach that FDA intends to use to evaluate the results of studies from which scientific conclusions can be drawn and rate the strength of the total body of publicly available evidence.

In a 2010 case before the U.S. District Court for the District of Columbia, a dietary supplement manufacturer challenged FDA's decisions regarding specific health claims.[42] At issue in this case were five proposed health claims describing a purported relationship between selenium and the risk of cancer. FDA had denied four of the claims outright, and it stated that it would exercise enforcement discretion regarding a modified version of the remaining claim. The manufacturer challenged the denial of the four claims and the modified claim, describing the modified claim as an imposition of an onerous, value-laden set of qualifications and as unreasonably long and burdensome for industry to include.[43] The court examined FDA's review of the scientific evidence, referencing procedures laid out in the agency's 2009 guidance document. The court questioned a number of FDA's conclusions and remanded the denied claims to the agency to reevaluate.[44] In addition, the court found FDA's modified disclaimer at odds with the First Amendment. The court stated that "FDA is obligated to at least consider the possibility of approving [the manufacturer's] proposed language with the addition of [a] 'short, succinct, and accurate disclaimer[].'"[45]

## APPENDIX II: OBJECTIVES, SCOPE, AND METHODOLOGY

We examined (1) the results of FDA's efforts to allow the use of qualified health claims on food, and its oversight of these claims and (2) what is known about consumers' understanding of qualified health claims on food. In addition, we examined industry's use of structure/function claims on food and FDA's oversight of these claims.

To examine the results of FDA's efforts to allow food companies to use qualified health claims on food, we assessed FDA data and documentation, including health claim guidance and petitions from industry. We reviewed petitions for qualified health claims and FDA's letters to petitioners stating that if a company used the submitted claim as modified by FDA, the agency would exercise its enforcement discretion and not take action against the claim. We also requested data on FDA resources devoted to qualified health claims and petitions. Although FDA does not maintain data on expenditures and staff charges that would allow it to provide the actual amount it expended on qualified health claims since 2000, FDA estimated that its Center for Food Safety and Applied Nutrition (CFSAN) expended at least $12.8 million since 2000, including the expenses associated with about 88.7 full-time equivalent staff, to administer health and qualified health claim activities for food labeling. This estimate does not represent FDA's total expenditures on activities related to health claims and qualified health claims for the period because FDA officials could not provide estimates or actual data on expenditures and staff years for other than CFSAN with any degree of confidence.

We reviewed the literature to identify the extent to which different health- and nutrient-related claims are used on food labels in grocery stores and interviewed industry associations and others regarding the marketing advantages and disadvantages of using qualified health and structure/function claims. We obtained data from articles by FDA and West Virginia University researchers to determine the use of qualified health and structure/function claims on food labels. We assessed the quality of the studies on which the articles were based and found them to be of sufficient quality for the purposes of this review.

To examine FDA's oversight of qualified health claims, we reviewed FDA guidance and related documents on inspection and enforcement of food labeling claims. We also reviewed those warning letters and enforcement actions that FDA officials were able to identify having taken in response to violations, since 2001, as well as the results of its January 2010 initiative to target oversight of labeling compliance. We assessed the reliability of FDA's data on petitions and health claims by (1) reviewing available information about the data and the system that produced them, (2) interviewing agency officials knowledgeable about the data, and (3) comparing the results with other sources of data for reasonableness. When we found inconsistencies in the data, we discussed them with agency officials and worked with them to correct the inconsistencies. We determined the data were sufficiently reliable for the purposes of this review.

To examine what is known about consumers' understanding of qualified health claims on conventional food labels, we conducted a literature search of various databases, such as PubMed, MEDLINE, and ProQuest, to identify relevant research studies; we also asked stakeholders we interviewed to recommend studies. Our selection criteria were that the studies be based on original research on consumers' understanding of qualified health claims and be less than 10 years old. We identified six research studies, conducted by FDA, the Federal Trade Commission (FTC), the International Food Information Council, and academia, that met these criteria. In addition, FDA and the International Food Information Council identified research studies they had conducted that also addressed consumers' understanding of other types of claims on food labels, including structure/function claims. These were the aforementioned International Food Information Council study and an additional FDA study. Two GAO social science analysts reviewed, assessed, and agreed on the quality of each of the studies we used and found that, while all the studies had some limitations, the limitations were not so great as to preclude the studies' use.

We identified stakeholders from health, consumer, and industry groups, as well as researchers, to represent the views of different sectors with regard to consumer perceptions and policy options related to qualified health claims and structure/function claims. We asked FDA and the National Academies' Institute of Medicine to comment on the list of stakeholders we developed to ensure fairness and balance. The final list included the following:

- *health associations:* the American Heart Association, American Dietetic Association, American Medical Association, and the American Society for Nutrition;
- *consumer groups:* the Consumer Federation of America and the Center for Science in the Public Interest;
- *an industry association:* the Grocery Manufacturers Association; and
- *researchers:* the International Food Information Council; FTC; FDA; and Neal Hooker, Chair of Food Marketing, St. Joseph's University, Philadelphia, Pennsylvania.

We also interviewed food labeling officials at agencies from the two largest exporters of agricultural products to the United States—Canada and the

European Union—to determine how these trading partners regulated qualified health and structure/function claims.

To examine industry's use of structure/function claims and FDA's oversight of these claims, we reviewed studies by FDA and researchers at West Virginia University and discussed the results with FDA compliance officials. We also obtained copies of FDA's warning letters and discussed those and FDA oversight activities. We met with FTC officials and compared and contrasted FDA's and FTC's responsibilities and authorities with respect to food labeling and advertising claims, and examined actions taken by FTC against a company for deceptive structure/function claims. We also met with FTC officials to discuss the commission's interactions with FDA on overseeing the proper use of claims on food labeling and advertising.

We conducted this performance audit from January 2010 through December 2010 in accordance with generally accepted government auditing standards. Those standards required that we plan and perform the audit to obtain sufficient, appropriate evidence to provide a reasonable basis for our findings and conclusions based on our audit objectives. We believe that the evidence obtained provides a reasonable basis for our findings and conclusions based on our audit objectives.

# Appendix III: Types of Claims FDA Recognizes, and Industry May Use, on Food

FDA recognizes several types claims that are used on food labels, including the following.

## Health Claims

- *Health claims* may be authorized by FDA when it determines that there is significant scientific agreement among qualified experts that a totality of publicly available scientific evidence supports the claim, as provided under the Nutrition Labeling and Education Act of 1990. For example, "Diets low in sodium may reduce the risk of high blood pressure, a disease associated with many factors" is one such health claim.

- *Health claims* may also be based on an authoritative statement of a scientific body of the U.S. government or the National Academy of Sciences, as provided under the Food and Drug Administration Modernization Act of 1997;[46] such claims may be used 120 days after notification to FDA. One such claim states: "Diets rich in whole grain foods and other plant foods and low in total fat, saturated fat, and cholesterol may reduce the risk of heart disease and some cancers."
- *Qualified health claims* are health claims to which FDA has added qualifying language to characterize the strength and limitations of the scientific evidence in support of the claim's potentially misleading nature, as described in FDA's 2003 *Consumer Health Information for Better Nutrition Initiative* guidance. One such claim states: "Very limited and preliminary scientific research suggests that eating one-half to one cup of tomatoes and/or tomato sauce a week may reduce the risk of prostate cancer. FDA concludes that there is little scientific evidence supporting this claim."

## Structure/Function Claims

*Structure/function claims* describe the role of a nutrient to affect a bodily structure (e.g., bones) or bodily function (e.g., digestion), or characterize the mechanism by which a nutrient acts to maintain such structure or function. For example, "Calcium helps build strong bones." FDA is not required by statute to and does not review or approve these claims.

## Nutrient Content Claims

- *Nutrient content claims* characterize the level of a nutrient in a food using terms such as "free," "high," and "low" (e.g., "high in vitamin C"), or compare the level of a nutrient in a food with that of another food, using terms such as "more," "reduced," and "lite," (e.g., "contains 10 percent more of the Daily Value for fiber than white bread," "reduced fat—50 percent less fat than our regular brownies," or "lite cheese cake, 1/3 fewer calories and 50 percent less fat than our regular cheesecake"), as provided under the Nutrition Labeling and Education Act of 1990.

- *Nutrient content claims* may be based on an authoritative statement of a scientific body of the U.S. government or the National Academy of Sciences, as provided under the Food and Drug Administration Modernization Act of 1997. One such claim is "good source of choline," which must reference the amount or percentage of the daily value of choline in each serving. (Choline is an essential nutrient usually grouped within the B-complex vitamins.)

## Implied Nutrient Content Claims

- Claims about a food or ingredient that suggest that the nutrient is absent or present in a certain amount or claims about a food that suggest a food may be useful in maintaining healthy dietary practices and that are made with an explicit claim (e.g., "healthy, contains 3 grams of fat") are implied claims.
- Claims that a food contains or is made with an ingredient that is known to contain a particular nutrient may be made if the product is "low" in or a "good source" of the nutrient associated with the claim (e.g., "good source of oat bran").
- Equivalence claims—"contains as much [nutrient] as a [food]"—may be made if both reference food and labeled food are a "good source" of a nutrient on a per serving basis (e.g., "contains as much vitamin C as an 8 ounce glass of orange juice").

The following label statements are generally *not* considered implied claims unless they are made in a nutrition context: (1) avoidance claims for religious, food intolerance, or other non-nutrition-related reasons (e.g., "100% milk free"); (2) statements about non-nutritive substances (e.g., "no artificial colors"); (3) added value statements (e.g. "made with real butter"); (4) statements of identity (e.g., "corn oil" or "corn oil margarine"); and (5) special dietary statements made in compliance with a specific provision of Part 105 of Title 21 of the Code of Federal Regulations.

# APPENDIX IV: HEALTH CLAIMS THAT MAY BE USED ON FOOD LABELS

Table 5 provides information, including sample claim language, on the 12 health claims authorized for foods that meet certain requirements that are based on FDA's determination that there is significant scientific agreement among qualified experts that all publicly available scientific evidence supports the claims. Table 6 provides information, including the required wording, on the 5 health claims for food that are based on an authoritative statement of scientific body of the U.S. government or the National Academy of Sciences.

**Table 5. Twelve Health Claims Based on Significant Scientific Agreement That May Be Used on Food Labels**

| Food/disease risk | Claim requirements | Model claim |
|---|---|---|
| Calcium/ osteoporosis | The claim must<br>"make clear that adequate calcium or calcium and vitamin D intake throughout life in a healthful diet are essential to reducing osteoporosis risk."<br>"not imply that adequate calcium or adequate calcium and vitamin D intake is the only recognized risk factor for the development of osteoporosis."<br>"not attribute any reduction in risk of osteoporosis to maintaining an adequate dietary calcium or dietary calcium and vitamin D intake throughout life." | *Calcium and osteoporosis health claim*:<br>"Adequate calcium throughout life, as part of a well-balanced diet, may reduce the risk of osteoporosis" or "Adequate calcium as part of a healthful diet, along with physical activity, may reduce the risk of osteoporosisin later life."<br>*Calcium, vitamin D, and osteoporosis*:<br>"Adequate calcium and vitamin D throughout life, as part of a well-balanced diet, may reduce the risk of osteoporosis" or "Adequate calcium and vitamin D throughout life, along with physical activity, may reduce the risk of osteoporosis in later life." |
| Sodium/ hypertension | *Required terms*:<br>"Sodium," "high blood pressure."<br>Includes physician statement (Individuals with high blood pressure should consult their physicians) if claim defines high or normal blood pressure | Diets low in sodium may reduce the risk of high blood pressure, a disease associated with many factors. |
| Dietary lipids/ cancer | *Required terms*:<br>"Total fat" or "fat."<br>"Some types of cancers" or "some cancers."<br>Does not specify types of fats or fatty acids that may be related to risk of cancer. | Development of cancer depends on many factors. A diet low in total fat may reduce the risk of some cancers. |

| Food/disease risk | Claim requirements | Model claim |
|---|---|---|
| Dietary saturated fat and cholesterol/ coronary heart disease | *Required terms*:<br>"Coronary heart disease" or "heart disease."<br>Includes physician statement (individuals with elevated blood total—or LDL—cholesterol should consult their physicians) if claim defines high or normal blood total—and LDL— cholesterol. | While many factors affect heart disease, diets low in saturated fat and cholesterol may reduce the risk of this disease. |
| Fiber-containing grain products, fruits, and vegetables/ cancer | *Required terms*:<br>"Fiber," "dietary fiber," or "total dietary fiber."<br>"Some types of cancer" or "some cancers."<br>Does not specify types of dietary fiber that may be related to risk of cancer. | Low-fat diets rich in fiber-containing grain products, fruits, and vegetables may reduce the risk of some types of cancer, a disease associated with many factors. |
| Fruits, vegetables and grain products that contain fiber, particularly soluble fiber/ coronary heart disease | *Required terms*:<br>"Fiber," "dietary fiber," "some types of dietary fiber," "some dietary fibers," or "some fibers."<br>"Saturated fat" and "cholesterol."<br>"Heart disease" or "coronary heart disease."<br>Includes physician statement ("Individuals with elevated blood total—or LDL—cholesterol should consult their physicians") if claim defines high or normal blood total—and LDL— cholesterol. | Diets low in saturated fat and cholesterol and rich in fruits, vegetables, and grain products that contain some types of dietary fiber, particularly soluble fiber, may reduce the risk of heart disease, a disease associated with many factors. |
| Fruits and vegetables/cancer | *Required terms*:<br>"Fiber," "dietary fiber," or "total dietary fiber."<br>"Total fat" or "fat."<br>"Some types of cancer" or "some cancers."<br>Characterizes fruits and vegetables as "Foods that are low in fat and may contain vitamin A, vitamin C, and dietary fiber."<br>Characterizes specific food as a "good source" of one or more of the following: dietary fiber, vitamin A, or vitamin C.<br>Does not specify types of fats or fatty acids or types of dietary fiber that may be related to risk of cancer. | Low-fat diets rich in fruits and vegetables (foods that are low in fat and may contain dietary fiber, vitamin A, or vitamin C) may reduce the risk of some types of cancer, a disease associated with many factors. Broccoli is high in vitamins A and C, and it is a good source of dietary fiber. |
| Folate/neural tube defects | *Required terms*:<br>Terms that specify the relationship (e.g., women who are capable of becoming pregnant and who consume adequate amounts of folate) "folate," "folic acid," | Healthful diets with adequate folate may reduce a woman's risk of having a child with a brain or spinal cord defect. |

**Table 5. (Continued)**

| Food/disease risk | Claim requirements | Model claim |
|---|---|---|
| | "folacin," "folate, a B vitamin," "folic acid, a B vitamin," "folacin, a B vitamin," "neural tube defects," "birth defects, spinal bifida, or anencephaly," "birth defects of the brain or spinal cord—anencephaly or spinal bifida," "spinal bifida or anencephaly, birth defects of the brain or spinal cord."<br>Must also include information on the multifactorial nature of neural tube defects, and the safe upper limit of daily intake. | |
| Dietary noncariogenic carbohydrate sweeteners/dental caries | *Required terms*:<br>"Does not promote," "may reduce the risk of,"<br>"useful [or is useful] in not promoting," or<br>"expressly [or is expressly] for not promoting" dental caries.<br>"Dental caries," or "tooth decay."<br>"Sugar alcohol" or "sugar lcohols" or the name or names of the sugar alcohols; or D-tagatose, or sucralose.<br>Note: D-tagatose may be identified as "tagatose."<br>When the substance that is the subject of the claim is a noncariogenic sugar (i.e., D-tagatose) the claim shall identify the substance as a sugar that, unlike other sugars, does not promote the development of dental caries.<br>Includes statement that frequent between-meal consumption of foods high in sugars and starches can promote tooth decay.<br>Packages with less than 15 square inches of surface area available for labeling may use a shortened claim. | Full claim: Frequent between-meal consumption of foods high in sugars and starches promotes tooth decay. The sugar alcohols in [name of food] do not promote tooth decay.<br>Shortened claim (on small packages only):<br>Does not promote tooth decay. |
| Soluble fiber from certain foods/ coronary heart disease | *Required terms*:<br>"Heart disease" or "coronary heart disease."<br>"Saturated fat" and "cholesterol."<br>In specifying the substance the claim uses the term "soluble fiber"qualified by the name of the eligible source of the soluble fiber, which is either whole oat or barley or psyllium seed husk. Claim specifies the daily dietary intake of the soluble fiber source necessary to | Soluble fiber from foods such as [*name of soluble fiber source, and, if desired, name of food product*], as part of a diet low in saturated fat and cholesterol, may reduce the risk of heart disease. A serving of [*name of food product*] supplies __ grams of the [necessary daily dietary intake for the benefit] soluble fiber from [*name of soluble fiber source*] necessary per day to have this effect. |

| Food/disease risk | Claim requirements | Model claim |
|---|---|---|
| | reduce the risk of coronary heart disease. Claim specifies the amount of soluble fiber in one serving of the product.<br>Additional required label statement: Foods bearing a psyllium seed husk health claim must also bear a label statement concerning the need to consume them with adequate amounts of fluids; e.g., "NOTICE: This food should be eaten with at least a full glass of liquid. Eating this product without enough liquid may cause choking. Do not eat this product if you have difficulty in swallowing." | |
| Soy protein/ coronary heart disease | *Required terms*:<br>"Heart disease" or "coronary heart disease."<br>"Soy protein."<br>"Saturated fat" and "cholesterol."<br>Claim specifies daily dietary intake levels of soy protein associated with reduced risk.<br>Claim specifies amount of soy protein in a serving of food. | (1) Twenty-five grams of soy protein a day, as part of a [diet] low in saturated fat and cholesterol, may reduce the risk of heart disease. A serving of [*name of food*] supplies __ grams of soy protein.<br>(2) Diets low in saturated fat and cholesterol that include 25 grams of soy protein a day may reduce the risk of heart disease. One serving of [*name of food*] provides __ grams of soy protein. |
| Plant sterol/stanol esters/coronary heart disease | *Required terms*:<br>"May" or "might" reduce the risk of coronary heart disease.<br>"Heart disease" or "coronary heart disease,"<br>"Plant sterol esters" or "plant stanol esters"; except "vegetable oil" may replace the term "plant" if vegetable oil is the sole source of the sterol/ stanol ester.<br>Claim specifies plant sterol/stanol esters are part of a diet low in saturated fat and cholesterol.<br>Claim does not attribute any degree of coronary heart disease risk reduction.<br>Claim specifies the daily dietary intake of plant sterol or stanol esters necessary to reduce coronary heart disease risk, and the amount provided per serving.<br>Claim specifies that plant sterol or stanol esters should be consumed with two different meals each a day. | (1) Foods containing at least 0.65 grams per [serving] of vegetable oil sterol esters, eaten twice a day with meals for a daily total intake of least 1.3 grams, as part of a diet low in saturated fat and cholesterol, may reduce the risk of heart disease. A serving of [*name of food*] supplies __ grams of vegetable oil sterol esters.<br>(2) Diets low in saturated fat and cholesterol that include two servings of foods that provide a daily total of at least 3.4 grams of plant stanol esters in two meals may reduce the risk of heart disease. A serving of [*name of food*] supplies __ grams of plant stanol esters. |

Source: FDA.

**Table 6. Five Health Claims Based on Authoritative Statements That May Be Used on Food Labels**

| Food/disease risk | Food requirements | Required wording of health claim |
|---|---|---|
| Whole grain foods/ reduced risk of heart disease and certain cancers | Contains 51 percent or more whole grain ingredients by weight per referenced amount customarily consumed (RACC), and Dietary fiber content at least:<br>3.0 grams (g) per RACC of 55 g<br>2.8 g per RACC of 50 g<br>2.5 g per RACC of 45 g<br>1.7 g per RACC of 35 g<br>Low fat | Diets rich in whole grain foods and other plant foods and low in total fat, saturated fat, and cholesterol may reduce the risk of heart disease and some cancers. |
| Potassium/reduced risk of high blood pressure and stroke | Good source of potassium.<br>Low sodium.<br>Low total fat.<br>Low saturated fat.<br>Low cholesterol. | Diets containing foods that are a good source of potassium and that are low in sodium may reduce the risk of high blood pressure and stroke. |
| Fluoridated water/ reduced risk of dental caries | Bottled water meeting the standards of identity and quality set forth in 21 CFR 165.110<br>Meet all general requirements for health claims in 21 CFR 101.14) with the exception of the minimum nutrient contribution (21 CFR 101.14(e)(6))<br>Total fluoride<br>>0.6 to 1.0 mg/L.<br>Excluding bottled water products specifically marketed for use by infants. | Drinking fluoridated water may reduce the risk of [dental caries, or tooth decay]. |
| Saturated fat, cholesterol, and trans fat/ reduced risk of heart disease. | Low saturated fat.<br>Low cholesterol.<br>Bear quantitative *trans* fat labeling.<br>Contain less than 0.5 g *trans* fat per RACC.<br>Contain less than 6.5 g total fat. | Diets low in saturated fat and cholesterol, and as low as possible in *trans* fat, may reduce the risk of heart disease. |
| Substitution of saturated fat with unsaturated fatty acids/ reduced risk of heart disease | Vegetable oils, spreads, and shortenings that have a total unsaturated fat content of 80 percent or more of total fat. | Replacing saturated fat with similar amounts of unsaturated fats may reduce the risk of heart disease. To achieve this benefit, total daily calories should not increase. |

Source: FDA.

# APPENDIX V: WARNING LETTERS ISSUED FOR CLAIM VIOLATIONS ON FOOD LABELS AND WEB SITES, DECEMBER 2009 THROUGH FEBRUARY 2010

FDA launched an initiative to address claim violations in food labeling in late 2009 and early 2010 in which it issued 17 warning letters and an untitled letter informing companies that the claims on their product labels and/or Web sites violated FDA's food labeling statutes and regulations. Two of the 17 warning letters were for qualified health claims violations and are summarized in table 4 of this report. The remaining 15 warning letters and the untitled letter are summarized in table 7.

**Table 7. Other Warning Letters and the Untitled Letter FDA Issued for Labeling Claim Violations, December 2009 through February. 2010**

| Action | Company (date of letter) | Product and/or product line | Violation(s) |
|---|---|---|---|
| **Warning letters** | | | |
| | Nestle USA (Dec. 4, 2009) | Juicy Juice Brain Development Fruit Juice Beverage Apple) | The product makes claims such as "no sugar added," which are not allowed on products intended for children under 2 years of age because appropriate dietary levels have not been established for children in this age range. |
| | | Juicy Juice All-Natural 100% Juice Orange Tangerine; Juicy Juice All-Natural 100% Juice Grape | The product labels imply that the products are 100 percent juice when they are actually juice blends with added flavors. |
| | Nestle Nutrition (Feb. 22, 2010) | Gerber's 2nd Foods Carrots | The product makes claims such as "healthy," "excellent source of . . . vitamin A," and "no added sugar," which are not allowed on products intended for children under 2 years of age because appropriate dietary levels have not been established for children in this age range. |
| | | Gerber Graduates Fruit Puffs line | The product makes claims such as "good source of iron, zinc, and vitamin E," which are not allowed on products intended for children under 2 years of age because appropriate dietary levels have not been established for children in this age range. |

**Table 7. (Continued)**

| Action | Company (date of letter) | Product and/or product line | Violation(s) |
|---|---|---|---|
| | Beech-nut Nutrition Corporation (Feb. 22, 2010) | Beechnut DHA Plus line of products | The product makes claims on its Web site such as “no added refined sugar” and "plus vitamins and minerals," which are not allowed on products intended for children under 2 years of age because appropriate dietary levels have not been established for children in this age range. |
| | | Beechnut Whole Grain Oatmeal with mixed fruit | The product makes claims such as “low sodium,” “plus fiber,” and “plus vitamins & minerals,” which are not allowed on products intended for children under 2 years of age because appropriate dietary levels have not been established for children in this age range. |
| | Dreyers Grand Ice Cream, Inc. (Feb. 22, 2010) | Nestle Drumstick Classic Vanilla Fudge | The front panel shows that the product has no trans fat, but it does not have a disclosure statement to alert consumers that the product has significant levels of saturated fat and total fat. |
| | | Dreyers Dibs Bite Sized Ice Cream Snacks Vanilla Ice Cream with Nestle Crunch Coating | The front panel shows that the product has no trans fat, but it doesn't have a disclosure statement to alert consumers that the product has significant levels of saturated fat and total fat. |
| | First Juice, Inc. (Feb. 22, 2010) | Organic Fruit and Veggie Juice Beverage products | The product make claims such as “plus calcium,” and "50% less sugar," which are not allowed on products intended for children under 2 years of age because appropriate dietary levels have not been established for children in this age range. |
| | | Purple carrot products | The products make claims that the products will “reduce the risk of cancer and stroke,” This claim has not been authorized by FDA for use on food products. |
| | Gorton’s, Inc. (Feb. 22, 2010) | Gorton's Beer Battered Crispy Battered Fish Fillets | The front panel shows that the product has no trans fat, but it does not have a disclosure statement to alert consumers that the product has significant levels of sodium, saturated fat, and total fat. |

| Action | Company (date of letter) | Product and/or product line | Violation(s) |
|---|---|---|---|
| | Guangzhou Yong Want Foods Ltd. (Feb. 22, 2010) | Baby Mum-Mum Original Selected Superior Rice Rusks | The product makes claims such as "low in fat," and "no added fats for oils," which are not allowed on products intended for children under 2 years of age because appropriate dietary levels have not been established for children in this age range. |
| | Ken's Foods, Inc. (Feb. 22, 2010) | Ken's Healthy Options™ Dressings Parmesan & Peppercorn; Sweet Vidalia® Onion Vinaigrette; Raspberry Walnut | The product makes claims such as "Healthy Options," but has more fat than is allowed in products labeled as "healthy." |
| | PBM Products, LLC (Feb. 22, 2010) | Parent's Choice Little Puffs Plus Calcium Blueberry Naturally Flavored product; Parent's Choice Little Puffs Made With Whole Grains Peach-Mango Naturally Flavored product. | The product makes claims such as "plus calcium," which are not allowed on products intended for children under 2 yrs of age because appropriate dietary levels have not been established for children in this age range. |
| | Pompeian, Inc. (Feb. 22, 2010) | Pompeian Imported Extra Light Olive Oil | The product makes nutrient content claims such as "light," and "high in good monounsaturated fat," but does not meet the requirement to make these claims. |
| | Redco Foods, Inc. (Feb. 22, 2010) | Salada Naturally Decaffeinated Green Tea | The product makes claims that it will treat, prevent, or cure diseases such as Alzheimer's disease, rheumatism, and cancer. These types of claims are not allowed on food products. |
| | | | The product makes claims on the product Web site that it is effective in the prevention of cardiovascular disease, but this claim has been not authorized for this product. |
| | | | The product makes claims such as "fortified with antioxidants," but the claim does not meet the requirements of the antioxidant regulation. |
| | Schwan's Consumer Brands (Feb. 22, 2010) | Mrs. Smith's Classic Coconut Custard Pie | The front panel shows that the product has no trans fat, but it does not have a disclosure statement to alert consumers that the product has significant levels of saturated fat and total fat. |
| | Spectrum Organic Products, Inc. (Feb. 22, 2010) | Organic All Vegetable Shortening | The front panel shows that the product has no trans fat, but it does not have a disclosure statement to alert consumers that the product has |

**Table 7. (Continued)**

| Action | Company (date of letter) | Product and/or product line | Violation(s) |
|---|---|---|---|
| | | | significant levels of saturated fat and total fat. The product makes nutrient content claims such as "cholesterol free," "less saturated fat than butter," and "good source of . . . monounsaturated fat" but does not meet the legal requirements to make these claims. |
| | Sunsweet Growers, Inc. (Feb. 22, 2010) | Antioxidant Blend Dried Fruit Mix | The product makes claims such as "full of nutritious antioxidants," but the claim does not meet the requirements of the antioxidant regulation. |
| | POM Wonderful (Feb. 23, 2010) | POM Wonderful 100% Pomegranate Juice | The product makes claims that it will treat, prevent, or cure diseases such as hypertension, diabetes, and cancer. These types of claims are not allowed on food products. |
| **Untitled letter** | | | |
| | Nature's Path Foods, Inc. (Feb. 22, 2010) | Organic Flax Plus Multibran Cereal | The product label includes the nutrient claim, "excellent source of Omega-3+," which has not been approved for use on food products. |

Source: GAO analysis of FDA information.

## End Notes

[1] In this report we use the word "allow" to describe FDA's assertion to companies that the agency will consider exercising its discretion to not take enforcement action against the company for making a certain health claim so long as that claim is made in accordance with criteria provided by FDA.

[2] GAO, *Food Labeling: FDA Needs to Better Leverage Resources, Improve Oversight, and Effectively Use Available Data to Help Consumers Select Healthy Foods*, GAO-08-597 (Washington, D.C.: Sept. 9, 2008).

[3] Consolidated Appropriations Act, 2008, Pub. L. No. 110-16 1 § 4, 121 Stat. 1844, 1846 (2007) (referencing explanatory statement at 153 Cong. Rec. H15742, H15765).

[4] Manufacturers of dietary supplements making structure/function claims must notify FDA no later than 30 days after the first marketing of the dietary supplement with such a claim statement.

[5] The Federal Food, Drug, and Cosmetic Act, as amended by the FDA Food Safety Modernization Act on January 4, 2011, authorized FDA to order a mandatory recall if it determines there is a reasonable probability that an article of food is (1) adulterated or (2) misbranded with regard to a major food allergen, and the use of or exposure to the food will cause serious adverse health consequences or death in humans or animals.

[6] GAO, *High-Risk Series: An Update,* GAO-07-310 (Washington, D.C.: January 2007).

[7] GAO-08-597.

[8] Five of the 16 petitions were initially submitted as health claim petitions under the significant scientific agreement standards but were converted to qualified health claim petitions at the request of the petitioners. All 16 petitions proposed claims for use on food labels, although some were for use on both food and dietary supplement labels.

[9] FDA noted in its comments that in each of the last 10 years, it has accomplished its work related to health claims and qualified health claims by supporting an average 8.1 FTEs (at 1,700 hours per FTE) per year at a cost of approximately $1 million per year.

[10] According to the FLAPS study, FDA analyzed the 2005 AC Nielsen Strategic Planner database to identify a sample of 3,000 U.S. supermarkets with over $2 million in annual revenue, across all geographic areas. Only products available in at least 2 percent of the stores in the retail market were considered. The data can be generalized to 80 to 85 percent (plus or minus 3 percent) of the retail food sales.

[11] Paula Fitzgerald Bone and Karen Russo France, "Qualified Health Claims on Package Labels," *Journal of Public Policy & Marketing*, Vol. 28 (2) Fall 2009, 253–258.

[12] The International Food Information Council is a nonprofit, nonpartisan communications organization whose mission is to effectively communicate science-based information about food safety and nutrition to health professionals, government officials, consumers, and others.

[13] Conrad J. Choinière and Linda Verrill, FDA, *Experimental Study of Qualified Health Claims: Consumer Inferences about Monounsaturated Fatty Acids from Olive Oil, EPA and DHA Omega-3 Fatty Acids, and Green Tea* (November 2009).

[14] Stanol esters are a group of chemical compounds that reduce low-density lipoprotein (LDL) cholesterol. They are found naturally occurring in small quantities in fruits, vegetables, nuts, seeds, whole grain, legumes, and vegetable oils, and are added to certain food for their health benefits.

[15] Wendy Reinhardt Kapsak, David Schmidt, Nancy M. Childs, John Meunier, and Christy White; "Consumer Perceptions of Graded, Graphic and Text Label Presentations for Qualified Health Claims," *Critical Reviews in Food Science and Nutrition* (March 2008) vol. 48, issue 3.

[16] R. Dennis Murphy, FTC, *Consumer Perceptions of Qualified Health Claims in Advertising*, Washington, DC (July 2005).

[17] Remarks of David Vladeck, Director, FTC Bureau of Consumer Protection, before the National Advertising Division Annual Conference, New York, N.Y., Oct. 5, 2009.

[18] Chung-Tung Jordin Lin, FDA "How Do Consumers Interpret Health Messages on Food Labels?" *Nutrition Today,* Vol. 43., No. 6 (November/December 2008).

[19] Neal Hooker and Ratapol Teratanavat, "Dissecting Qualified Health Claims: Evidence From Experimental Studies," *Critical Reviews in Food Science and Nutrition,* Vol. 48, Iss. 2 (February 2008).

[20] Paula Fitzgerald Bone and Karen Russo France, "Qualified Health Claims on Package Labels," *Journal of Public Policy & Marketing,* Vol. 28 (2), Fall 2009, 253–258.

[21] FDA, Center for Food Safety and Applied Nutrition, *Guidance for Industry: Substantiation of Dietary Supplement Claims Made Under Section 403(r) (6) of the Federal Food, Drug, and Cosmetic Act.*

[22] FDA informed us that its data system for tracking warning letters could not be searched for structure/function claims on food. FDA officials were able to identify only the two letters discussed herein.

[23] FTC does not use the term "structure/function claim," but FDA confirmed that the claim met its understanding of a structure/function claim.

[24] GAO, *Food Labeling: FDA Needs to Better Leverage Resources, Improve Oversight, and Effectively Use Available Data to Help Consumers Select Healthy Foods*, GAO-08-597 (Washington, D.C.: Sept. 9, 2008).

[25] Pub. L. No. 101-535, 104 Stat. 2353.

[26] 21 U.S.C. §§ 301-399a.

[27] In addition, a claim may be authorized if a relevant scientific body of the federal government or the National Academies, or a subdivision thereof, has published an authoritative statement, currently in effect, about the relationship between a nutrient and a disease or health-related condition to which the claim refers.

[28] 56 Fed. Reg. 60,537, Nov. 27, 1991.

[29] Dietary Supplement Act of 1992, Pub. L. No. 102-571, tit. II § 202, 106 Stat. 4500.

[30] 58 Fed. Reg. 2478, Jan. 6, 1993.

[31] 59 Fed. Reg. 395, Jan. 4, 1994.

[32] 164 F.3d 650 (D.C. Cir. 1999).

[33] In addition, the court found that the Administrative Procedure Act requires that FDA give some "definitional content to the phrase 'significant scientific agreement,'" either in regulation or on a case-by-case basis so that the regulated class can "perceive the principles which are guiding agency action." 164 F.3d at 661. FDA subsequently provided guidance describing the meaning of the phrase.

[34] 65 Fed. Reg. 59,855, Oct. 6, 2000.

[35] 67 Fed. Reg. 78,002, Dec. 20, 2002.

[36] 68 Fed. Reg. 41,387, July 11, 2003.

[37] *Whitaker v. Thompson*, 248 F. Supp. 2d 1 (D.D.C. 2002). See also *Pearson v. Shalala*, 130 F. Supp. 2d 105 (D.D.C. 2001).

[38] Those qualifying statements being (1) "although there is scientific evidence supporting the claim, the evidence is not conclusive"; (2) "some scientific evidence suggests . . . however, FDA has determined that this evidence is limited and not conclusive"; and (3) "very little and preliminary scientific research suggests . . . FDA concludes that there is little scientific evidence supporting this claim." FDA, *Interim Procedures for Qualified Health Claims in the Labeling of Conventional Human Foods and Human Dietary Supplements* (July 10, 2003).

[39] 68 Fed. Reg. 66,040, Nov. 25, 2003.

[40] FDA, Center for Food Safety and Applied Nutrition, *Guidance for Industry: Evidence- Based Review System for the Scientific Evaluation of Health Claims* (January 2009).

[41] FDA, Center for Food Safety and Applied Nutrition, *Guidance for Industry and FDA: Interim Evidence-Based Ranking System for Scientific Data* (July 10, 2003).

[42] *Alliance for Natural Health US v. Sebelius*, 714 F. Supp. 2d 48 (D.D.C. 2010).

[43] 714 F. Supp. 2d at 58.

[44] The court directed FDA reevaluate one of the claims and draft one or more disclaimers. For the remaining three denied claims, the court directed FDA to either draft disclaimers to accompany the manufacturer's claims, or set forth the empirical evidence that any disclaimer would fail to correct the claims' purported misleading effects. 714 F. Supp. 2d at 72.

[45] 714 F. Supp. 2d at 71 (quoting *Pearson v. Shalala*).

[46] Pub. L.105-115 § 303, 111 Stat. 2296.

In: Food Labeling: FDA Protections...
Editors: J. K. Scott and M. I. Hayes

ISBN: 978-1-61942-759-4


*Chapter 2*

# COUNTRY-OF-ORIGIN LABELING FOR FOODS

***Remy Jurenas***

## SUMMARY

Many retail food stores are now required to inform consumers about the country of origin of fresh fruits and vegetables, seafood, peanuts, pecans, macadamia nuts, ginseng, and ground and muscle cuts of beef, pork, lamb, chicken, and goat. The rules are required by the 2002 farm bill (P.L. 107- 171) as amended by the 2008 farm bill (P.L. 110-246). Other U.S. laws have required such labeling, but only for imported food products already pre-packaged for consumers.

Both the authorization and implementation of country-of-origin labeling (COOL) by the U.S. Department of Agriculture's Agricultural Marketing Service have not been without controversy. Much attention has focused on the labeling rules that now apply to meat and meat product imports. A number of leading agricultural and food industry groups continue to oppose COOL as costly and unnecessary. They and some major food and livestock exporters to the United States (e.g., Canada and Mexico) also view the new requirement as trade-distorting. Others, including some cattle and consumer groups, maintain that Americans want and deserve to know the origin of their foods, and that many U.S. trading partners have their own, equally restrictive import labeling requirements.

Obama Administration officials announced in February 2009 that they would allow the final rule on COOL, published just before the end of the Bush Administration on January 15, 2009, to take effect as planned on March 16, 2009. However, the Secretary of Agriculture also urged affected industries to adopt—voluntarily—several additional changes that, the Obama Administration asserts, would provide more useful origin information to consumers and also would more closely adhere to the intent of the COOL law.

Retail compliance with COOL requirements appears to have proceeded reasonably well. To address identified labeling problems, observers have called for additional outreach to retailers to help them better understand what is required and the steps they can take to improve compliance.

The most significant issue that has arisen to date is the November 2009 decision by Canada and Mexico to challenge COOL rules and the "voluntary suggestions" using the World Trade Organization's (WTO's) trade dispute resolution process. Both countries argue that COOL has a trade-distorting impact by reducing the shipment of their cattle and hogs to the U.S. market, as U.S. livestock market participants began to make adjustments in anticipation of new meat labeling rules. They also argue that COOL rules violate trade rules that the United States agreed to under the WTO and the North American Free Trade Agreement. Responses to this development reflect the heated debate seen earlier among key players in the livestock sector. U.S. meatpackers and processors support Canada's and Mexico's position that COOL violates U.S. trade obligations. Some cattle producer groups argue that COOL is consistent with U.S. commitments and does not discriminate between imported and domestic beef. Twenty-five Senators have expressed their support for COOL, noting that other countries (including Canada and Mexico) require country-of-origin information to be provided to consumers.

The 111$^{th}$ Congress is considering legislation that would expand COOL labeling requirements to cover more food products. H.R. 2749, the House-passed food safety bill, would expand such labeling to apply to all processed foods and to other agricultural commodities not now covered by the farm bill and other statutory provisions. The Senate-reported companion bill (S. 510) does not include a comparable provision, but pertinent amendments may be offered during floor debate. Separately, S. 1783 would require retailers to implement COOL for dairy products.

## RECENT DEVELOPMENTS

On June 16, 2010, the European Parliament voted to accept draft food labeling legislation that would expand food country-of-origin labeling (COOL) requirements throughout the European Union (EU). Mandatory COOL is already required for beef, honey, olive oil, and fresh fruits and vegetables. The rules now would also apply to all meat, poultry, dairy products, and other single- ingredient foods. Also, country-of-origin information for meat, poultry, and fish used as an ingredient in a processed food product would need to be shown. The measure will next be considered by the EU's 27 member governments early next year, before returning back to the European Parliament for final consideration. If then adopted, the measure would take effect in three to five years.[1]

On May 10, 2010, the World Trade Organization's (WTO) director named three members of the dispute settlement panel that will consider Canada's and Mexico's complaint that the U.S. law and rules on mandatory COOL are inconsistent with its several WTO-related trade commitments that the United States has entered into (see "WTO Case Brought Against COOL by Canada and Mexico").

Half of surveyed consumers reported that COOL is extremely or very important when making decisions to purchase fresh meat, fish, fruits, or vegetables, according to a March 2010 food survey conducted by consulting firm Deloitte. Survey results showing that consumers want to know more about food nutrition, quality, and safety also revealed that 45% would like to learn on a website the country of origin for all ingredients in a packaged or bottled food product.[2]

## BACKGROUND

Since the 1930s, U.S. tariff law has required almost all imports to carry labels so that the "ultimate purchaser," usually the retail consumer, can determine their country of origin. However, certain products, including a number of agricultural commodities in their "natural" state such as meats, fruits and vegetables, were excluded. For almost as many decades, various farm and consumer groups have pressed Congress to end one or more of these exceptions, arguing that U.S. consumers have a right to know where all of their food comes from and that, given a choice, they would purchase the domestic version. This would strengthen demand and prices for U.S. farmers and ranchers, it was argued.

Opponents of ending these exceptions to country-of-origin labeling (COOL) contended that there was little or no real evidence that consumers want such information and that industry compliance costs would far outweigh any potential benefits to producers or consumers. Such opponents, including other farm and food marketing groups, argued that mandatory COOL for meats, produce, or other agricultural commodities was a form of protectionism that would undermine U.S. efforts to reduce foreign barriers to trade in the global economy. COOL supporters countered that it was unfair to exempt agricultural commodities from the labeling requirements that U.S. importers of almost all other products already must meet, and that major U.S. trading partners impose their own COOL requirements for imported meats, produce, and other foods.

With passage of the 2002 farm bill (P.L. 107-171, § 10816), retail-level COOL was to become mandatory for fresh fruits and vegetables, beef, pork, lamb, seafood, and peanuts, starting September 30, 2004. Continuing controversy over the new requirements within the food and agricultural industry itself led Congress to postpone full implementation. The FY2004 omnibus appropriations act (P.L. 108-199) postponed COOL—except for seafood—until September 30, 2006; the FY2006 agriculture appropriation (P.L. 109-97) further postponed it until September 30, 2008.[3]

During deliberations on a new omnibus farm bill in 2007 and 2008, those affected by COOL reached consensus on a series of amendments intended to ease what many of them viewed as some of the more onerous provisions of the 2002 COOL law. Modified were provisions dealing with record-keeping requirements, the factors to be considered for labeling U.S. and non-U.S. origin products, and penalties for noncompliance. These amendments were incorporated into the final farm bill (P.L. 110-246, § 11002). The enacted 2008 bill required that COOL take effect on September 30, 2008, and added goat meat, chicken, macadamia nuts, pecans, and ginseng as commodities covered by mandatory COOL.

Final rules to fully implement the COOL requirements were published by the U.S. Department of Agriculture's (USDA's) Agricultural Marketing Service (AMS) during the final days of the Bush Administration in January 2009. Obama Administration officials announced in February 2009 that they would allow the final rules to take effect as planned on March 16, 2009. However, they also urged affected industries to adopt—voluntarily—several additional changes that, the Administration asserts, would provide more specific origin information to consumers and more closely adhere to the intent of the COOL law.

Major U.S. trading partners, including Canada and Mexico, which are challenging COOL using the WTO's dispute settlement process, continue to closely watch implementation developments. So are lawmakers in the 111$^{th}$ Congress, where additional legislation to expand COOL' s coverage to include other food products is being considered.

# OTHER LAWS WITH LABELING PROVISIONS

## Tariff Act

Under §304 of the Tariff Act of 1930, as amended (19 U.S.C. 1304), every imported item must be conspicuously and indelibly marked in English to indicate to the "ultimate purchaser" its country of origin. The U.S. Customs and Border Protection generally defines the "ultimate purchaser" as the last U.S. person to receive the article in the form in which it was imported. So, articles arriving at the U.S. border in retail-ready packages—including food products, such as a can of Danish ham, or a bottle of Italian olive oil—must carry such a mark. However, if the article is destined for a U.S. processor where it will undergo "substantial transformation," the processor is considered the ultimate purchaser. Over the years, numerous technical rulings by Customs have determined what is, or is not, considered "substantial transformation," depending upon the item in question.

The law has authorized exceptions to labeling requirements, including articles on a so-called "J List," named for § 1 304(a)(3)(J) of the statute. This empowered the Secretary of the Treasury to exempt classes of items that were "imported in substantial quantities during the five-year period immediately preceding January 1, 1937, and were not required during such period to be marked to indicate their origin." Among the items placed on the J List were specified agricultural products including "natural products, such as vegetables, fruits, nuts, berries, and live or dead animals, fish and birds; all the foregoing which are in their natural state or not advanced in any manner further than is necessary for their safe transportation."[4] Although J List items themselves have been exempt from the labeling requirements, § 304 of the 1930 act has required that their "immediate container" (essentially, the box they came in) have country-of-origin labels. But, for example, when Mexican tomatoes or

Chilean grapes are sold unpackaged at retail in a store bin, country labeling had not been required by the Tariff Act.

## Meat and Poultry Products Inspection Acts

USDA's Food Safety and Inspection Service (FSIS) is required to ensure the safety and proper labeling of most meat and poultry products, including imports, under the Federal Meat Inspection Act, as amended (21 U.S.C. 601 et seq.), and the Poultry Products Inspection Act, as amended (21 U.S.C. 451 et seq.). Regulations issued under these laws have required that country of origin appear in English on immediate containers of all meat and poultry products entering the United States (9 C.F.R. 327.14 and 9 C.F.R. 381.205). Only plants in countries certified by USDA to have inspection systems equivalent to those of the United States are eligible to export products to the United States.

All individual, retail-ready packages of imported meat products (for example, canned hams or packages of salami) have had to carry such labeling. Imported bulk products, such as carcasses, carcass parts, or large containers of meat or poultry destined for U.S. plants for further processing, also have had to bear country-of-origin marks. However, once these non-retail items have entered the country, the federal meat inspection law has deemed them to be domestic products. When they are further processed in a domestic, FSIS-inspected meat or poultry establishment—which has been considered the ultimate purchaser for purposes of country-oforigin labeling—F SIS no longer requires such labeling on either the new product or its container. FSIS has considered even minimal processing, such as cutting a larger piece of meat into smaller pieces or grinding it for hamburger, enough of a transformation so that country markings are no longer necessary.

Meat and poultry product imports must comply not only with the meat and poultry inspection laws and rules but also with Tariff Act labeling regulations. Because Customs generally requires that imports undergo more extensive changes (i.e., "substantial transformation") than required by USDA to avoid the need for labeling, a potential for conflict has existed between the two requirements.

### Federal Food, Drug, and Cosmetic Act

Foods other than meat and poultry are regulated by the U.S. Department of Health and Human Services' Food and Drug Administration (FDA), primarily under the Federal Food, Drug, and Cosmetic Act (FFDCA; 21 U.S.C. 301 et seq.). This act does not expressly require COOL for foods. Section 403(e) of the FFDCA does regard a *packaged* food to be misbranded if it lacks a label containing the name and place of business of the manufacturer, packer, or distributor (among other ways a food can be misbranded). However, this name and place of business is not an indicator of the origin of the product itself. (See "Legislation" for a proposal to amend the FFDCA to penalize sellers if they do not label the country of origin of processed foods and of non-processed foods that are not covered by farm bill provisions.)

## IMPLEMENTATION OF FARM BILL COOL REQUIREMENTS

The COOL provisions of the 2002 and 2008 farm bills do not change the requirements of the Tariff Act or the food safety inspection statutes; rather, they amend the Agricultural Marketing Act (AMA) of 1946 (7 U.S.C. 1621 note). USDA's Agricultural Marketing Service (AMS) administers most AMA-authorized programs, including COOL.[5] AMS published a final rule to implement COOL for all covered commodities on January 15, 2009, which took effect on March 16, 2009.[6] This final rule, issued during the closing days of the Bush Administration, replaces both the April 4, 2005, interim final rule for seafood (see footnote 3), and the August 1, 2008, interim final rule for all other covered commodities.[7]

The COOL rule was re-examined in early 2009 by newly confirmed Secretary of Agriculture Tom Vilsack to comply with an Obama White House directive that all agencies review recent regulations issued by the outgoing Administration. After this re-examination, Secretary Vilsack announced, on February 20, 2009, that the regulations would take effect as planned on March 16. However, the Secretary's announcement included two significant provisos. First, he urged affected industries to voluntarily adopt several suggested labeling changes in order to provide more useful information to consumers than the rule itself might imply, and to better meet congressional intent. Second, he stated that USDA would closely monitor industry compliance to determine whether "additional rulemaking may be necessary to provide consumers with adequate information."[8]

The Secretary's "suggestions for voluntary action" were detailed in a February 20, 2009, letter sent to industry representatives. They deal with the treatment of meat products with multiple countries of origin, exemptions in the rules for processed products, and time allowances provided to ground meat manufacturers regarding their inventory.[9] These are further detailed below, in the appropriate sections.

During FY2009, state agencies conducted about 12,000 surveillance reviews (compared to 2,000 in FY2008) to ensure retail compliance with COOL requirements. The significant increase reflects the expansion of COOL's coverage from only fish and shellfish to include more commodities (see list below in "Key Provisions"). Agencies of all 50 states now perform this function under cooperative agreements signed with the AMS, which included participating in training sessions on their enforcement responsibilities. AMS also has developed educational and outreach activities for state cooperators, retailers, suppliers, and other interested parties. During FY2010, AMS plans to develop an expanded real-time database to track the findings of federal- state retail reviews, enforcement actions taken, and other information viewed as critical to COOL operations.[10]

Audits to gauge retail store efforts to meet COOL rules for vegetables and fruit have found the absence of any label to be the main noncompliance issue. Other issues discovered include nonspecific labeling (e.g., not identifying the origin of "locally grown" produce) and conflicting or inaccurate declarations as to the origin of the produce displayed. Industry representatives suggest that these issues can be resolved by helping retailers better understand the details laid out in COOL regulations, and encouraging them to be more diligent in labeling local produce. AMS currently is focused on using additional store reviews to look for possible differences between states in compliance levels, and analyzing whether supermarket chains operate differently than single-store retailers in meeting COOL rules.[11]

## Key Provisions

Mandatory country-of-origin labeling:

- *applies* to ground and muscle cuts of beef, lamb, and pork, farm-raised and wild fish and shellfish, peanuts, "perishable agricultural commodities" as defined by the Perishable Agricultural Commodities Act (PACA, i.e., fresh and frozen fruits and vegetables), goat meat, chicken, pecans, macadamia nuts, and ginseng (these are referred to as "covered commodities").[12]

- *exempts* these items if they are an ingredient in a processed food.
- *covers* only PACA-regulated retailers (those purchasing at least $230,000 a year in fresh fruits and vegetables), and requires them to inform consumers of origin "by means of a label, stamp, mark, placard, or other clear and visible sign on the covered commodity or on the package, display, holding unit, or bin containing the commodity at the final point of sale."
- *exempts* from these labeling requirements such "food service establishments" as restaurants, cafeterias, bars, and similar facilities that prepare and sell foods to the public.

## Defining Origin

In designating country of origin, difficulties arise when products—particularly meats—are produced in multiple countries. For example, beef might be from an animal that was born and fed in Canada, but slaughtered and processed in the United States. Likewise, products from several different countries often are mixed, such as for ground beef. For covered red meats and chicken, the COOL law:

- *permits* the U.S. origin label to be used only on items from animals that were exclusively born, raised, and slaughtered in the United States, with an exception for those animals present here before July 15, 2008;
- *permits* meats or chicken with multiple countries of origin to be labeled as being from all of the countries in which the animals may have been born, raised, or slaughtered;
- *requires* meat or chicken from animals imported for immediate U.S. slaughter to be labeled as from both the country the animal came from and the United States;
- *requires* products from animals not born, raised, or slaughtered in the United States to be labeled with their correct country(ies) of origin; and

- *requires*, for ground meat and chicken products, that the label list all countries of origin, or all "reasonably possible" countries of origin.

The meat labeling requirements have proven to be among the most complex and controversial areas of rulemaking, in large part because of the steps that U.S. feeding operations and packing plants must adopt to segregate, hold, and slaughter foreign-origin livestock from U.S. livestock. After issuance of the interim rules in August 2008, many retailers and meat processors reportedly had planned to use the "catch-all" label (see second bullet, above) on as much meat as possible— even products that would qualify for the U.S.-only label, because it was both permitted and the easiest requirement to meet. COOL supporters objected that the label would be overused, undermining the whole intent of COOL (i.e., to distinguish between U.S. and non-U.S. meats).[13] In an effort to balance the concerns of both sides, USDA issued a statement attempting to clarify its August 2008 interim rule, stating that meats derived from both U.S.- and non-U.S.-origin animals may carry a mixed-origin claim (e.g., "Product of U.S., Canada, and Mexico"), but that the mixed-origin label cannot be used if only U.S.-origin meat was produced on a production day.[14]

The final (January 2009) rule attempts to further clarify the "multiple countries of origin" language. For example, muscle cut products of exclusively U.S. origin along with those from foreign-born animals, if commingled on a single production day, can continue to qualify for a combined U.S. and non-U.S. label. "It was never the intent of the Agency [AMS] for the majority of product eligible to bear a U.S. origin declaration to bear a multiple origin destination. The Agency made additional modifications for clarity," AMS stated in material accompanying the rule.[15]

The clarifying change failed to mollify some producer groups, who have continued to view this portion of the rule as a "loophole that would allow meat packers to use a multiple countries, or NAFTA [North American Free Trade Agreement] label, rather than labeling U.S. products as products of the United States. This is misleading to consumers," stated National Farmers Union (NFU) President Tom Buis.[16]

In his February 20, 2009 letter, Secretary Vilsack asked industry representatives to voluntarily provide additional information. He stated that:

> processors should voluntarily include information about what production step occurred in each country when multiple countries appear on the label. For example, animals born and raised in Country

X and slaughtered in Country Y might be labeled as "Born and Raised in Country X and Slaughtered in Country Y." Animals born in Country X but raised and slaughtered in Country Y might be labeled as "Born in Country X and Raised and Slaughtered in Country Y."

The Vilsack letter also noted that the final rule allows a label for ground meat to bear the name of a country even if the meat from that country was not present in a processor's inventory in the preceding 60-day period. Noting that this allows for labeling this product "in a way that does not clearly indicate [its] country of origin," the Secretary asked processors to reduce this time allowance to 10 days, stating that this "would enhance the credibility of the label." The Vilsack letter is widely viewed as an effort to address the concerns of COOL adherents without reopening the rule and thereby attracting renewed criticism from the meat industry and U.S. trading partners.

For perishable agricultural commodities, ginseng, peanuts, pecans, and macadamia nuts, retailers may only claim U.S. origin if they were exclusively produced in the United States. However, a U.S. state, region, or locality designation is a sufficient U.S. identifier (e.g., Idaho potatoes). For farm-raised fish and shellfish, a U.S.-labeled product must be derived exclusively from fish or shellfish hatched, raised, harvested, and processed in the United States; wild fish and shellfish must be derived exclusively from those either harvested in U.S. waters or by a U.S. flagged vessel, and processed in the United States or on a U.S. vessel. Also, labels must differentiate between wild and farm-raised seafood.

## Coverage

Consumers may not find country-of-origin labels on much more of the food they buy, due to COOL's statutory and regulatory exemptions. First, as noted, all restaurants and other food service providers are exempt, as are all retail grocery stores that buy less than $230,000 a year in fresh fruits and vegetables. Second, "processed food items" derived from the covered commodities are exempt, and USDA, in its final rule, defined this term broadly (at 7 C.F.R. § 65.220). Essentially, any time a covered commodity is subjected to a change that alters its basic character, it is considered to be processed. Although adding salt, water, or sugar do not, under USDA's definition, change the basic character, virtually any sort of cooking, curing, or mixing apparently does. For example, roasting a peanut or pecan, mixing peas with carrots, or breading a piece of meat or chicken, all count as processing. As a result, only about 30% of the U.S. beef supply, 11% of all pork, 39% of chicken, and 40% of all fruit and vegetable supplies may be covered by COOL requirements at the retail level. [17] Whole peanuts are almost always purchased in roasted form, which will not have to be labeled. Some critics are arguing that AMS overstepped its authority, and congressional intent, by excepting such minimally processed commodities.

AMS had countered that in fact many imported items still must carry COOL under provisions of the Tariff Act of 1930. "For example, while a bag of frozen peas and carrots is considered a processed food item under the COOL final rule, if the peas and carrots are of foreign origin, the Tariff Act requires that the country of origin be marked on the bag," AMS argued, citing similar regulatory situations for roasted nuts and for a variety of seafood items.[18]

In his February 20, 2009 letter, however, Secretary Vilsack acknowledged that the "processed foods" definition in the final rule "may be too broadly drafted. Even if products are subject to curing, smoking, broiling, grilling, or steaming, voluntary labeling would be appropriate," he wrote.

## Record-Keeping, Verification, and Penalties

The COOL law prohibits USDA from using a mandatory animal identification (ID) system,[19] but the original 2002 version stated that the Secretary "may require that any person that prepares, stores, handles, or distributes a covered commodity for retail sale maintain a verifiable record-keeping audit trail that will permit the Secretary to verify compliance." Verification immediately became one of the most contentious issues, particularly for livestock producers, in part because of the potential complications and costs to affected industries of tracking animals and their products from birth through retail sale. Producers of the plant-based commodities, as well as food retailers and others, also expressed concern about the cost and difficulty of maintaining records for commodities that are highly fungible and often widely sourced. The 2008 law eased these requirements somewhat by stating that USDA "may conduct an audit of any person that prepares, stores, handles, or distributes a covered commodity" in order to verify compliance. Such persons must provide verification, but USDA may not ask for any additional records beyond those maintained "in the course of the normal conduct of business."

In its final rule, AMS stated that covered persons generally would have to keep records for one year that can identify both the immediate previous source and the immediate subsequent recipient of a covered commodity; certain exceptions are provided for pre-labeled products. Also, a slaughter facility can accept a producer affidavit as sufficient evidence for animal origin claims.

Also, potential fines for willful noncompliance are set for retailers and other persons at no more than $1,000 per violation. The 2002 law had set the fine at no more than $10,000 (and for retailers only), but the 2008 farm bill lowered this amount.

# ECONOMIC AND TRADE ISSUES

## Costs and Benefits

COOL supporters argue that a number of studies show that consumers want country-of-origin labeling and would pay extra for it. Analysis accompanying USDA's interim and final rules concluded that, while benefits are difficult to quantify, it appears they will be small and accrue mainly to consumers who desire such information. A Colorado State University economist suggested that consumers might be willing to pay a premium for "COOL meat" from the United States, but only if they perceive U.S. meat to be safer and of higher quality than foreign meat.[20] USDA earlier had estimated that purchases of (i.e., demand for) covered commodities would have to increase by between 1% and 5% for benefits to cover COOL costs, but added that such increases were not anticipated. Data from several economic modeling studies of COOL impacts appear to fall within this range.[21] Another research paper found that demand for domestic apples would need to increase by a range of 3% to 7% and for domestic tomatoes by 8% to 22% for COOL to increase total economic welfare in these markets.[22]

Critics of mandatory COOL have argued that large compliance costs will more than offset any consumer benefits. USDA's analysis of its final rule estimates that first-year implementation costs to be approximately $2.6 billion for those affected. Of the total, each commodity producer would bear an average estimated cost of $370, intermediary firms (such as wholesalers or processors) $48,219 each, and retailers $254,685 each. The USDA analysis also includes estimates of record- keeping costs and of food sector economic losses due to the rule.

## North American Livestock Trade

### *Overview*

With implementation now well underway, foreign suppliers—particularly Canada and Mexico— have questioned the trade legality of mandatory COOL. They have claimed that the August 2008 publication of the interim rule had an immediate impact in beginning to alter normal trade patterns and caused large financial losses. The initial focus of these concerns was on livestock (i.e., cattle and hogs, and their products).

For background, the animal products industries have become increasingly integrated across all three North American countries over the last two decades, particularly after NAFTA took effect in 1994 and, before that, the Canada-U.S. Free Trade Agreement in 1988. These agreements, along with the global Uruguay Round Agreements under the WTO, by helping to reduce tariff and nontariff barriers to trade, have enabled animals or their products to move across borders more freely, based on market demands. For example, in the pork industry, the Canadians tended toward breeding and farrowing small pigs, which in turn were shipped to the United States, where access to large supplies of grain made it more economical to feed them to slaughter weight.[23]

However, a number of animal health and other incidents have disrupted this market integration from time to time. The most significant event was the discovery of *bovine spongiform encephalopathy* (BSE) in 2003, first in Canada and later in the United States, which halted most cross-border movement of cattle from Canada to the United States from mid-2003 through mid- 2005. The predominance of BSE (mad cow) cases in Canada rather than in the United States may have contributed to wider support for the mandatory COOL law, some analysts believe, although government officials assert that both countries now have strong, scientifically defensible safeguards in place to ensure that BSE is controlled and that its infectious agent does not enter the human food supply.

### *U.S. Livestock Imports*

Almost all U.S. live cattle imports come from Canada and Mexico; all live hog imports come from Canada. After U.S. cattle and beef markets adjusted to the discovery of BSE in Canada's cattle herds in mid-2003, total cattle imports from both countries rose steadily each year from 2004 through 2007, reaching almost 2.5 million head (Figure 1). Similarly, annual hog imports from Canada almost doubled over the last several years, from 5.3 million head in 2001 to 10.0 million in 2007 (Figure 2).[24]

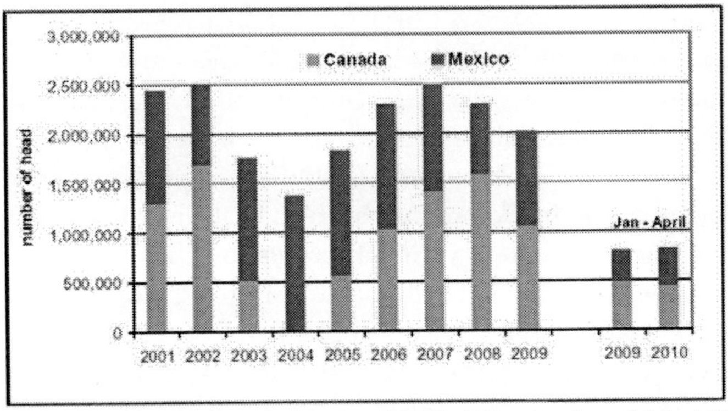


Source: USDA, Economic Research Service (ERS), "Livestock and Meat Trade Data – Cattle"

Figure 1. U.S. Cattle Imports from Canada and Mexico.

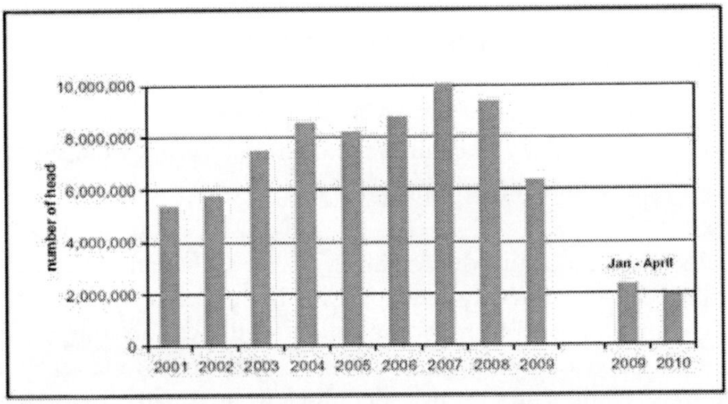


Source: USDA, ERS, "Livestock and Meat Trade Data – Hogs".

Figure 2. U.S. Hog Imports from Canada.

## Cattle

U.S. imports of feeder cattle (those destined for feedlots) and of slaughter-ready cattle (steers and heifers) from both Canada and Mexico declined on average 10% each year from 2007 to 2009. Imports totaled 2.0 million head in 2009, compared to 2.5 million head in 2007 (Figure 1). About 80% of the cattle imported from Canada are immediately slaughtered; all Mexican cattle enter to be grazed or fed. According to USDA, these imports have accounted for an estimated 10% of the U.S. cattle supply in recent years.

From July 2008 through June 2009, the period during which the U.S. beef sector began to prepare for and to implement COOL, U.S. cattle imports from both countries totaled almost 2.3 million head. This was 25% below the 3.0 million head imported during the same period in 2007/2008.

Some analyses attribute the import decline in the first several months of this period to COOL but differ on the extent that currency exchange rates may have contributed to this development. CattleFax, an industry-funded data and analysis service based in Colorado, observed that the 2008 decline in cattle imports were due to mandatory COOL regulations, and that imports would "face a big wild card in 2009" for the same reason.[25] Livestock sector analysts with the Chicago Mercantile Exchange (CME), examining cattle import trends through year-end 2008, commented that the COOL law "has been quite effective, if you measure effectiveness by the degree to which it has been able to stifle cattle trade in North America." They wrote that reductions in imports from both Mexico and Canada "came at a time when a significant devaluation in the value of the Peso and Canadian dollar normally would have been conducive of increased imports from these two countries. Under normal circumstances, one would expect cattle imports to actually increase rather than be cut by almost 40%."[26]

Late in 2008, however, USDA's Economic Research Service (ERS) suggested that the currency exchange factor may be somewhat more involved. The decline in Canadian cattle imports:

> coincided with the rapid depreciation of the Canadian dollar, making Canadian cattle relatively cheaper in U.S. dollar terms, but also making Canadian beef more competitive on the export market. Feeder cattle have been remaining in Canada. ... Imports of slaughter steers and heifers from Canada also declined dramatically in September, driven by the same exchange rate conditions affecting feeder cattle. Additionally, the past increase in Canadian exports of feeder cattle would reduce the current supply of fed cattle in Canada to be marketed or exported, lowering the number of Canadian cattle sent to the United States for slaughter.[27]

Other factors also have contributed to the decline in cattle imports since mid-2008. The drop in Mexican cattle shipments to the United States in the fourth quarter of 2008 compared to the same period a year earlier was due to good pasture conditions in Mexico and ranchers' decisions to hold on longer to their herds. The year-on-year fall in cattle imports from both countries during 2009 reflected less demand for cattle to slaughter, as processors adjusted to reduced domestic and foreign consumer demand for U.S. beef due to the economic recession (Figure 3).

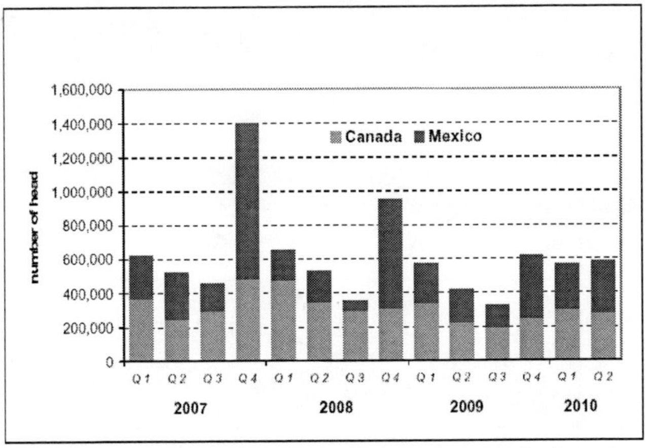


Source: Derived by CRS from USDA, Agricultural Marketing Service (AMS) published reports

Figure 3. U.S. Cattle Imports from Canada and Mexico (quarterly, 2007 to mid-2010).

## Hogs

U.S. imports of Canadian hogs (feeder pigs and hogs ready for slaughter) fell slightly in 2008 and more noticeably in 2009 (Figure 2). From July 2008 through June 2009 (the 12-month period that coincided with initial steps taken by U.S. hog producers and pork processors to prepare for COOL implementation), hog imports from Canada of almost 7.2 million head were 31% below the same period in 2007/2008 (10.5 million head).[28] Monthly hog imports have been lower on a year over year basis since mid-2008 (Figure 4).

Though developments in Canada's hog sector account in part for this drop, an early 2009 USDA analysis suggested that COOL's implementation likely "made U.S. swine finishers reluctant to import Canadian finishing animals, in light of some major U.S. packers' stated unwillingness to process Canadian-origin animals."[29] Another report suggested that COOL was affecting the U.S. hog sector, particularly in Iowa, as packers moved to process only U.S.-born hogs. With many Iowa producers operating finishing operations that source feeder pigs from Canada, a USDA document on COOL implementation cited that some producers' barns are "empty because of a lack of an assured outlet for slaughter hogs of mixed country of origin" (i.e., Product of Canada and United States). USDA also reported that some lenders were not extending credit to operations that finish mixed-origin pigs, and that lower prices at times were "being paid for mixed origin slaughter hogs compared to hogs of exclusively U.S. origin."[30]

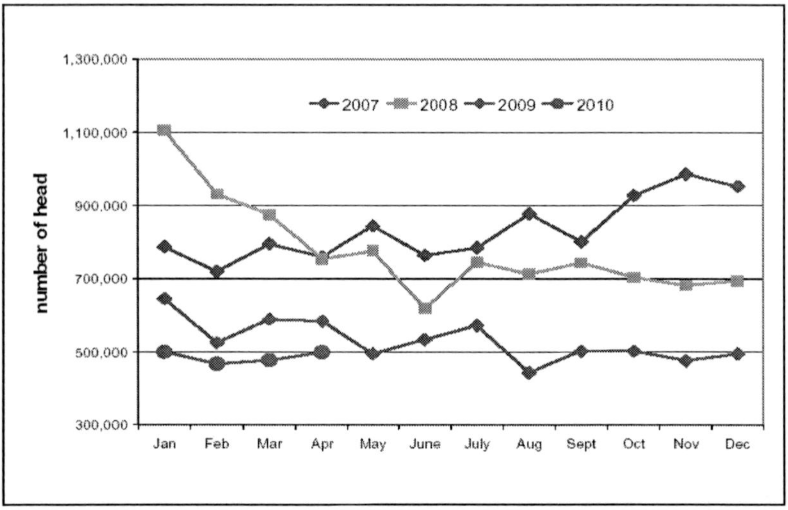


Source: USDA, ERS, "Livestock and Meat Trade Data – Hogs"

Figure 4. Monthly U.S. Hog Imports from Canada.

### *WTO Case Brought against COOL by Canada and Mexico*

Canada and Mexico are major suppliers of live cattle and hogs that are fed in U.S. facilities and/or processed into beef and pork in U.S. meat packing plants. As the U.S. meat processing sector geared up to implement COOL in mid-2008, Canadian and Mexican cattle and Canadian hog exports to the U.S. market initially declined. Both countries expressed concern about the adverse impacts this development had on their livestock sectors. Not satisfied with the outcome of two sets of consultations held with U.S. officials on their concerns (December 2008 and June 2009), both Canada and Mexico in November 2009 requested the establishment of a WTO dispute resolution panel to consider their case.

Both the Canadian and Mexican governments, in their requests for a panel, assert that COOL is inconsistent with several WTO-related trade commitments, including those providing that imports must be treated no less favorably than products of domestic origin; that laws on marks of origin should not damage imports, reduce their value, or unreasonably increase their cost; and that laws, rules, and procedures on country of origin should not "themselves" create or disrupt international trade, among other things.

In proceeding with a WTO case, Canadian officials stated that the COOL requirements are "so onerous" that when they were implemented, Canadian exporters of cattle and hogs were discriminated against in the U.S. market. The

Canadian beef and pork industries, led by the Canadian Cattlemen's Association (CCA) and the Canadian Pork Council, actively pushed their government to initiate a WTO challenge. The CCA argued that COOL cost its producers C$92 million (Canadian dollars) in losses over the two months following the publication of the interim rule in July 2008, and could cost C$500 million per year. CCA estimated that slaughter steers and heifers were losing C$90 per head, because U.S. meat establishments did not want to assume the increased costs of complying with new labeling requirements by segregating, holding, and then slaughtering Canadian cattle separately from U.S. cattle. The losses included lower prices for all Canadian cattle due to decreased U.S. demand, as well as the cost of shipping those that are sold further distances, to the fewer number of U.S. plants willing to take them. Canadian pork producers expressed similar concerns.[31]

U.S. officials regretted that the consultations did not resolve Canada's concerns, and stated their belief that U.S. implementation of COOL provides consumers with information that is consistent with WTO commitments. They noted that countries had agreed that COOL was legitimate policy long before the WTO was created, and that other countries also require goods to be labeled with their origin.[32]

In December 2009, 25 senators wrote to Obama Administration officials to express their support for implementing COOL according to congressional intent, stating that the program is nondiscriminatory in its requirements that both domestic and imported goods be labeled with their origin. They also noted that 45 other countries (including Canada and Mexico) implement programs that provide country of origin information to consumers.[33]

USTR's request for public comment on this pending WTO case generated responses that reflected the heated debate on mandatory COOL seen earlier among key players in the livestock sector. The American Meat Institute (AMI), representing U.S. meat processors and packers, stated that the U.S. law, in addition to violating WTO commitments, also violates NAFTA commitments. AMI argued that COOL discriminates against imports in favor of domestic meat.[34]

In opposition, the U.S. Cattlemen's Association (USCA) and the National Farmers Union argued that COOL is "fully consistent" with the General Agreement on Tariffs and Trade and the Agreement on Technical Barriers of Trade (key WTO commitments). Both stated that this U.S. law "does not discriminate between domestic and imported beef ... [and] operates neutrally in the market place," and noted that COOL does not impose any domestic content requirements (i.e., does not stipulate what share of value or quantity determines country of origin).[35]

Earlier, the National Cattlemen's Beef Association (NCBA) expressed concern that Canada's decision to pursue its case against U.S. COOL rules has the potential for retaliatory action to be taken against U.S. beef. It noted that "COOL has damaged critically important trading relationships [i.e., the import of Canadian and Mexican livestock, the value added as they pass through U.S. feedlots and are processed into meat, and the export of finished meat products back to Mexican and Canadian consumers], and is not putting additional money into the pockets of cattlemen."[36] The Ranchers-Cattlemen Action Legal Fund, United Stockgrowers of America (R- CALF USA), presented similar comments.[37]

The scope of the case that the WTO is now considering is not specific to livestock and their products. Presumably other covered commodities could be affected if the dispute panel rules against the U.S. COOL law.

## LEGISLATION

A number of lawmakers appear to agree with some industry groups' criticisms of mandatory COOL and could offer legislation to limit its scope and impacts. Other lawmakers remain strongly supportive of the new law and likely would oppose any significant rollback. Still others seek to bring more commodities under the scope of mandatory country-of-origin labeling.

Observers point out that the 2008 farm bill changes were intended to balance the concerns of both sides and, in effect, settle the longstanding controversy over requiring COOL. However, unfolding trade and market developments, including the WTO challenge initiated by Canada and Mexico and changes in import patterns, could alter the dynamics of any COOL debate in the 111$^{th}$ Congress and later.

### Expansion of COOL in Food Safety Measures

Provisions in one major food safety bill (H.R. 2749) would expand the scope of country-of-origin labeling to apply to all processed foods and other agricultural commodities not now covered under other statutory authorities. Earlier, during the 2008 debate over the safety of imported foods due in part to the scare over melamine in imports from China, some suggested that COOL be extended to additional products—a proposal included in an FDA reform bill and subsequent food safety measures.

Section 202 of H.R. 2749 (the House-passed food safety measure) would, by amending the Federal Food, Drug, and Cosmetic Act (FFDCA), significantly expand the country-of-origin food labeling requirement to apply to all processed foods, and to all non-processed foods not covered by the 2002 and 2008 farm bill provisions. As noted earlier, the FFDCA does not contain express COOL requirements for foods (or drugs). Failure to meet either of the following labeling requirements would be cause for FDA to deem such an incident a "misbranding" of the product, which would result in the imposition of fines and/or other penalties on the seller.[38] Specifically, a *processed food* would be "misbranded" if its label did not identify the country where final processing occurred. Relatedly, a *non-processed* food would be "misbranded" if its label failed to identify the country of origin. These requirements would take effect two years after enactment. Acknowledging that other federal laws already require COOL for food, this provision would direct FDA to develop its regulations to account for the labeling requirements already mandated by U.S. Customs and Border Protection for processed foods (see "Tariff Act") and by the Agricultural Marketing Service for non-processed foods (see "Key Provisions"). No comparable provision is in the Senate's food safety bill (S. 510) as reported by the Health, Education, Labor, and Pensions (HELP) Committee. Amendments to require COOL for additional products might be offered during Senate floor debate. An earlier version of the House bill (Section 133 of H.R. 759, 110th Congress) also would have required a manufacturer of a processed food to identify the country or countries of origin for each ingredient in the product, and an original packer of non- processed food to identify its country of origin, on their respective websites.

## Cool for Dairy Products

Separately, S. 1783 would amend the Agricultural Marketing Act of 1946 to provide for countryof-origin labeling for dairy products (fluid milk, cheese, yogurt, ice cream, butter, and other dairy products). Each retailer of dairy products would be required to designate the product's origin as (1) each country in which or from which the one or more dairy ingredients were produced or originated, and each country in which the product was processed; and (2) if a U.S.-produced dairy product, the state, region, or locality where the product was produced. Proponents argue that this "tool" would give consumers information to "quickly and easily identify" U.S.-made dairy products from milk that meets U.S. "high safety standards," in light of the 2008 discovery of melamine in Chinese dairy products.[39]

The National Milk Producers Federation, the largest trade association representing dairy producers, in late 2008 stated that it "does not desire to extend any COOL regulations to milk and milk products." It has not taken a position on this measure.[40] Observers point out that imports of fluid milk, cottage cheese, yogurt, and ice cream are minimal and that most imported cheeses and butter already are labeled with country of origin under other federal authority (see "Tariff Act"). The International Dairy Foods Association, a trade association representing dairy processors, opposes the bill. It argues that mandating COOL for dairy products and not for other processed foods "will reduce demand for [them] and encourage food manufacturers to substitute vegetable- based or other protein ingredients" instead.[41]

## End Notes

[1] Reuters, "UPDATE 1—EU Lawmakers Demand Stricter Food-Labelling Rules," June 16, 2010; *Food Business News*, "E.U. Votes in Favor of Uniform Food Labeling," July 6, 2010, p.30.

[2] *Feedstuffs*, "Consumers Seek Food Information," July 5, 2010, p. 7.

[3] An interim final rule for seafood COOL was published on October 5, 2004, and took effect April 4, 2005 (69 *Federal Register*, pp. 59708-59750).

[4] The J list is published in 19 C.F.R. 134.33, available at http://edocket.access19cfr134.33.htm.

[5] AMS maintains an extensive website on COOL, with links to implementing regulations, cost-benefit analysis, and other materials at http://www.ams.usda.gov /cool/.

[6] USDA, January 12, 2009, "USDA Issues Final Rule On Mandatory Country of Origin Labeling," available at http://www.usda.gov/wps/portal/usda/usdahome? printable=true&contentidonly=true&contentid=2009/01/0006.xml; 74 *Federal Register*, January 15, 2009, pp. 2658-2707. An AMS fact sheet on the final rule, including a summary of changes from the interim final rules and estimates on COOL implementation costs, is available at http://www.amsusda. gov/AMSv1.0/ getfile?dDocName=STELPRDC5074847.

[7] 73 *Federal Register*, August 1, 2008, pp. 45106-45151. AMS had indicated in August 2008 that it would not aggressively enforce the interim rule for six months (a period that, under the final rule as well, was to continue through March 2009) to give those affected more time to understand and fully comply with it.

[8] USDA, "Vilsack Announces Implementation of Country of Origin Labeling Law," February 20, 2009, available at http://www.usda.gov/wps/portal/usda/usda home? printable=true& contentidonly =true&contentid=2009/02/0045.xml.

[9] This letter is available at http://www.usda.gov/ documents/0220_Industry Letter COOL.pdf.

[10] USDA, FY2011 Budget Explanatory Notes for Committee on Appropriations for Agricultural Marketing Service, pp. 19-9, 19g-9, and 19-47, available at http://www.obpa.usda.gov/19ams2011notes.pdf.

[11] *Supermarket News*, "Simply COOL," May 24, 2010.

[12] A slightly different COOL requirement applies to packaged honey if it bears any official USDA certificate, mark, or statement with respect to quality and grade. It was added by

Section 10402 of the 2008 farm bill (P.L. 110-246) to the Agricultural Marketing Act, and took effect on October 6, 2009. For more information, see http://www.ams.usda.gov/AMSv1.0/ams.fetchTemplateData. do? startIndex=1&startIndex= 2&startIndex=1&startIndex=2&template=TemplateN&navID=ProcessedFVUpdates&right Nav1=&topNav=&leftNav=&page=ProcessedFVUpdates&resultType=&acct =procsdgrdcert.

[13] *Cattle Buyers Weekly*, August 4, 2008; and *Food Chemical News*, September 15, 2008.

[14] "Country of Origin Labeling (COOL) Frequently Asked Questions," September 26, 2008. Virtually all foreign live meat animals now come from either Canada or Mexico (see "North American Livestock Trade").

[15] USDA, AMS, January 12, 2009, fact sheet on the mandatory COOL final rule, available at http://www.ams.usda.gov/AMSv1.0/getfile?dDocName= STELPRDC 5074847.

[16] "NFU Statement: USDA Issues Final Rule for COOL," January 12, 2009, available at http://nfu.org/news/2009/01/12/nfu-statement-usda-issues-final-rule-for-cool. html.

[17] Percentages calculated by CRS based upon USDA estimates of retail-level COOL coverage in pounds, divided by total annual supply (USDA data on domestic production plus imports).

[18] USDA, AMS, "Frequently Asked Questions," January 12, 2009, available at http://www.ams.usda.gov/AMSv1.0/ getfile?dDocName=STELPRDC5074846.

[19] For information on this related issue, see CRS Report R40832, *Animal Identification and Traceability: Overview and Issues*, by Randy Schnepf.

[20] Wendy J. Umberger, "Will Consumers Pay a Premium for Country-of-Origin Labeled Meat?" *Choices*, 4th quarter 2004, available online at http://www. choices magazine.org/2004-4/cool/2004-4-04.htm.

[21] Gary W. Brewster et al., "Who Will Bear the Costs of Country-of-Origin Labeling?" (http://www.choicesmagazine.org/2004-4/cool/2004-4-02.htm); and Daniel D. Hanselka et al., "Demand Shifts in Beef Associated with Country-of-Origin Labeling to Minimize Losses in Social Welfare"(http://www.choices magazine.org/ 2004-4/cool/2004-4-03.htm), *Choices*, 4th quarter 2004.

[22] Alejandro Plastina and Konstantinos Giannakis, "Market and Welfare Effects of Mandatory Country-of-Origin Labeling in the U.S. Specialty Crops Sector," Selected Paper, American Agricultural Economics Association Annual Meeting, Portland, Oregon, 2007.

[23] See for example, USDA, Economic Research Service (ERS), *Market Integration of the North American Animal Products Complex* (LDP-M-131-01), May 2005.

[24] USDA, ERS, Livestock and Meat Trade Data series, available at http://www.ers.usda.gov/data/meattrade/.

[25] CME, "CattleFax Long Term Outlook Special Edition," December 12, 2008, p. 3.

[26] *CME Daily Livestock Report*, January 7, 2009.

[27] USDA, ERS. *Livestock, Dairy, and Poultry Outlook*, December 18, 2008, p. 8. ERS analysts point out that prior to 2008, the United States was easing the BSE-related restrictions on Canadian cattle imports; in November 2007, cattle over 30 months of age were again permitted to enter from Canada.

[28] Derived by CRS from AMS marketing report "Canadian Live Animal Imports by Destination," selected issues in 2008 and 2009.

[29] USDA, ERS, *Livestock, Dairy, and Poultry Outlook*, April 16, 2009, p. 4.

[30] CattleBuyers Weekly, "MCOOL Hurts Iowa Hog Finishers," April 27, 2009.

[31] Various trade publication reports, including *Cattle Buyers Weekly*, "MCOOL Has Cost Canadian Producers C$92M," December 8, 2008; *Agri-Pulse*, "COOL regulations create heartburn for Canadians," December 3, 2008; and *Washington Trade Daily*, December 2, 2008, pp. 3-4.

[32] U.S. Trade Representative, "Vilsack, Kirk Comment on Canadian Panel Request Regarding Country-of-Origin Labeling," October 7, 2009, available at http://www.ustr.gov/about-us/press-office/press-releases/2009/october/vilsackkirk-comment-canadian-panel-request-regard.

[33] Office of Senator Tim Johnson, "Johnson, Enzi Lead the Fight to Keep Congressional Intent on COOL as WTO Dispute Moves Forward," December 21, 2009, available athttp://johnson.senate.gov/public/index.cfm?p=PressReleases&ContentRecord_id= c48ed4be-eebd-43da-95a12b079437819a&ContentType_id=c3d73cfe-c14b-4676-96ed43a65aea57c0&Group_id=6ae28060-e7a2-46ba-bbab-cce51bb5cb91&MonthDisplay=12&YearDisplay=2009.

[34] AMI, "American Meat Institute Tells U.S. Trade Representative that Mandatory Country-of-Origin Labeling Violates International Trade Obligations," January 8, 2010, available at http://www.meatami.com/ht/display/ ReleaseDetails/i/56358.

[35] USCA, "USCA and Farmers Union Urge Vigorous COOL Defense," January 12, 2010, available at http://www.uscattlemen.org/TheNewsRoom/2010_News/1-12COOL defense.htm.

[36] NCBA, "NCBA Statement on Canadian WTO Complaint against U.S. COOL Law," October 7, 2009, available at http:// www.beefusa.org /NEWSNCBAStatementonCanadianWTOComplaintagainstUSCOOLLaw39616.aspx.

[37] R-CALF USA, "Canada, Mexico Have No Standing to Bring Complaint Against U.S. COOL Law," July 2, 2009, available at http://www.r-calfusa.com/news_ releases/2009/090702-canada.htm.

[38] The concept of misbranding is one of the basic components of the FFDCA, and persons who violate its prohibitions are subject to criminal and civil penalties, as well as injunctions and seizures of the misbranded product.

[39] Office of Senator Al Franken, "Franken, Feingold, and Brown Introduce Dairy COOL Act," October 14, 2009, http://franken.senate.gov/?p=press_release&id= 563.

[40] NMPF, "Comments on USDA's Interim Final Rule on Mandatory COOL," September 29, 2008, available athttp://www.nmpf.org/files/file/NMPF% 20COOL %20comments%2009-29-08.pdf; E-mail communication from Dana Brooks, NMPF, July 8, 2010.

[41] *Supermarket News*, "Dairy Group Opposes COOL Law," October 29, 2009, p. 19.

# Chapter Sources

Chapter 1 - This is an edited, reformatted and augmented version of a Government Accountability Office publication, GAO-11-102, dated January, 2011.

Chapter 2 - This is an edited, reformatted and augmented version of a Congressional Research Service publication, RS22955, dated July 15, 2010.

# INDEX

## N



## O



## P



## Q



## R



## S